The ★ Magic of Ceramics

With a Foreword by Bonnie J. Dunbar, Ph.D., NASA Astronaut

DAVID W. RICHERSON

Published by

THE AMERICAN CERAMIC SOCIETY
735 CERAMIC PLACE
WESTERVILLE, OH 43081

The American Ceramic Society
735 Ceramic Place
Westerville, Ohio 43081
www.ceramics.org

ISBN: 1-57498-050-5

Printed in Hong Kong.

04 03 02 01 00   5 4 3 2

Cover photos are courtesy of *Durel Corporation* (cell phone); *Active Control eXperts* (skier); *OSRAM SYLVANIA* (light bulb); *Carbonmedics, Inc.* (finger joint); *Utah Museum of Fine Arts* (vase); *Kurt Nassau* (ruby); *Saint Gobain/Norton* (inline skate); *Rado Watch Co., Ltd.* (watch); *Buffalo China,* (mug); *David Richerson* (golf club); *Sandia National Laboratories* (helmet); and *Summitville Tiles* (tile).

Library of Congress Cataloging-in-Publication Data

Richerson, David W., 1944 –
    The magic of ceramics / David W. Richerson; with a foreword by Bonnie J. Dunbar
        p.cm.
    Includes index.
    ISBN 1-57498-050-5
    1. Ceramics. I. Title

    TP807.R53 2000
    666--dc21   99-056648

For information on ordering titles published by the American Ceramic Society, or to request a publications catalog, please call 614/794-5890. Visit our online bookstore at <www.ceramics.org>.

Contents

Preface & Acknowledgments...iv

Foreword v

Introduction vi

CHAPTER ONE

Our Constant Companions...2

CHAPTER TWO

From Pottery to the Space Shuttle 16

CHAPTER THREE

The Beauty of Ceramics...36

CHAPTER FOUR

Ceramics and Light..60

CHAPTER FIVE

Amazing Strength and Stability 88

CHAPTER SIX

Ceramics and the Electronics Age...122

CHAPTER SEVEN

Piezo Power...148

CHAPTER EIGHT

Medical Miracles 168

CHAPTER NINE

Ceramics and the Modern Automobile...190

CHAPTER TEN

Heat Beaters...210

CHAPTER ELEVEN

The Hardest Materials in the Universe 238

CHAPTER TWELVE

Energy and Pollution Control...262

Conclusion 284

Index...286

Preface & Acknowledgments

The Magic of Ceramics was written as part of an educational outreach project of the American Ceramic Society (ACerS). ACerS, founded in 1898, scheduled a year of centennial activities in 1998 and 1999. As part of the celebration, the Education Committee created a museum exhibit to introduce the general public and students to the amazing uses of ceramics. This was the nucleus of *The Magic of Ceramics*.

The idea for *The Magic of Ceramics* was proposed to ACerS, but with special circumstances. The book would be different from the technical books that ACerS typically publishes; it would also be entertaining, colorful, and available to the general public. ACerS accepted this proposal. Mary Cassells and Sarah Godby provided the direction and support to make this book a reality.

I sincerely thank all companies and individuals who donated ceramic parts, photographs, and their time in offering information for the exhibit and the book: M. Reynolds, S. Morse, T. Johnson, B. Jones, C. Davis, D. Keck, D. Duke, E. Sturdevant and M. Cotton, *Corning, Inc.*; L. Vallon, S. Dunlap, T. Taglialavore, B. Licht, B. McEntire, D. Croucher, M. Leger, E. Levadnuk, D. Parker, C.L. Quackenbush, T. Leo and J. Caron, *Saint-Gobain/Norton*; F. Moore, *Norton Chemicals*; D. Reed, J. Mangels and J. Moskowitz, *Ceradyne, Inc.*; K. Inamori, Y. Hamano, T. Yamamoto, J. Scovie, E. Kraft, and D. Carruthers, *Kyocera Corp.*; J. Price and M. van Roode, *Solar Turbines*; and H. Coors, B. Seegmiller, K. Michas, and L. Sobel, *Coors Ceramics*.

R. DeWolf, *Thielsch Engineering*; J. Knickerbocker, *IBM*; A. Bogue, *Active Control eXperts (ACX)*; D. Cavenaugh and J. Greenwald, *Wilbanks*; M. Kokta, K. Heikkinen, and L. Rothrock, *Union Carbide Crystal Products*; M. and M. Kasprzyk, *INEX Inc.*; S. Limaye, *LoTEC*; T. Sweeting, *HI-TECH Ceramics*; C. Greskovich and B. Riedner, *General Electric Co.*; M. Savitz and R. Walecki, *AlliedSignal Ceramic Components*; M. Easley and J. Kidwell, *AlliedSignal Engines*; R. Schultze, *AlliedSignal Filters and Spark Plugs*; F. Luhrs, *Rauschert Industries*; J. Ormerod, *Group Arnold*; and M. Kreppel, *Saint-Gobain/Carborundum*.

B. Powell and A. Micheli, *General Motors*; J. Staggs, *General Magnetic Co.*; P.Barker, *ITT Automotive*; B. Knigga, *Delco Electronics*; C. Strug, *American Superconductor*; F. Kennard, *Delphi Electronic Systems*; G. Melde, *Siemens Power Corp.*; J.Wenkus, *Zirmat Corp.*; A. Hecker and G. Bandyopadhyay, *OSRAM SYLVANIA*; G. Harmon, *American Marine*; T. Brenner, *Den-tal-ez, Inc.*; A. Vohra, *DOE*; W. Paciorek and L. Spotten, *Durel Corp.*; D. Witter, *MEMC Southwest*; J. Estes, *Motorola*; B. Cuttler, *Sandia National Laboratory*; P. Martin, *Hewlett Packard*; P. Gwordz, *San Jose State University*; and F. Schmidt and M. Felt, *Crystal Systems*.

J. Sapin, *Rado Watch Co., Ltd.*; V. Adams and C.A. Forbes, *Siemens-Westinghouse Electric*; R. Wasowski, *Ferro Corp.*; J. Lynch, *General Ceramics*; K. Elder, *Westvaco Corp.*; Y. Manring and W. Gates, *ACerS Ross Coffin Purdy Museum of Ceramics*; M. Dowley, *Liconix*; C. Chandler and A. Feidisch, *Spectra Physics, Inc.*; D. Bacso, G. Cook, R. Snow and C. Glassy, *Edo Piezoelectric Ceramic Products*; D. Greenspan and P. Neilson, *US Biomaterials*; A. Compton and W. Wolf, *Owens Corning*; St. Bender, *the History Factory*; S. Nelson, S. Logiudice and K. Budd, *3M*; L. Aubry, *Selee Corp.*; and L. Berthiaume, *Saphikon Inc.*

L. George, *National Dental Association*; D. Braski, *Oak Ridge National Laboratory*; A. Khandkar and S. Elangovan, *Ceramatec*; J. Hinton and V. Irick, *Lanxide Corp.*; A. Griffin and C. Griffin, *Lone Peak Engineering, Inc.*; T. Yonushonis and R. Stafford, *Cummins*; E. Lassow and T. Wright, *Howmet Research Corp.*; J. Buckley, *NASA Langley*; D. Day, *University of Missouri-Rolla*; R. Stoddard, *3M Unitek*; E. Pope, *Solgene Therapeutics, LLC*; L. Hench and J. Leonan, *Imperial College*; K. George, *Dynatronics*; and G. Fischman, *U.S. Food and Drug Administration*.

D. Kingery, *University of Arizona*; D. and N. Ferguson; T. Garcia; E.McEndarfer, *Truman State University*; W. Bates; K. Nassau; Marianne Letasi, *Detroit Institute of Art*; S. Rossi-Wilcox, *Botanical Museum of Harvard University*; J. Thomas-Clark, *Corning Museum of Glass*; J. Clottes, *International Committee on Rock Art, Foix, France*; J. Lever; and Count Robert Bégouën.

I am grateful to Angie Eagan who took my sketches and created illustrations; to J. Taylor, *Tile Heritage Foundation*, who provided slides of decorative tiles; to D. Whitehouse, *Corning Museum of Glass*, who offered suggestions of glass art to use as illustrations; to Ruth Butler, editor, *Ceramics Monthly*; and to W. South and D. Carroll, *Utah Museum of Fine Art*, for helping select items from the collection.

My special thanks to Alexis Clare, Margaret Carney, Jim Jacobs, Donaree Neville, Al and Barbara Kuipers, Michael Anne Richerson, and Jennifer Richerson for reading portions of the manuscript and providing suggestions. I am especially indebted to copy editor Susan Blake.

The last, but not least, person I want to thank is Mark Glasper, ACerS Director of Communications. He is an unsung hero in ACerS education efforts. The traveling museum exhibit would not have happened without his dedication and help.

DAVID W. RICHERSON

Foreword

I was born and raised in the Yakima Valley of eastern Washington state. My parents had homesteaded there in 1948, creating a ranch and farm in the middle of largely unirrigated, unpopulated land punctuated with sagebrush. We had no neighbors that I could see to the north, at the base of the Rattlesnake Mountains, and few to the south, east, or west. A two-lane dirt road ran by our house, which was later covered with gravel. During the summer months, after our chores were done, I would ask for my parents' permission to walk up and down that road for a certain distance to look for agates. But I collected more than agates; I collected all manner of interesting rocks, including chunks of quartz, mica, feldspar, and granite.

At night, there were no outdoor lights to diffuse the brightness of the stars. In fact, the Milky Way was a very large stripe across the sky. In the fall of 1957, my parents and I lay on the grass and looked for the first satellite launched into space—a Russian silver sphere called "Sputnik." It was during that time that I formed my own dreams to fly in space. Little did I realize at the time that my rock collection and flying in space would have something in common: they were both dependent upon chemical compositions, which we call "ceramics."

How these two apparently diverse worlds are joined is eloquently explained in *The Magic of Ceramics* by David W. Richerson. This wonderfully unique and readable book describes how humans have taken the rocks around us and, through chemistry, heat, and advanced technology, made them into glass, fiber optics, electronic and computer components, motor parts, tennis rackets, art work—in fact, the core of today's civilized technological society.

Materials have been referred to as the "enabling technology" of all other new engineering endeavors. Within this realm fall metals, organics, and ceramics. Readers may not be familiar with the breadth of "ceramics," but they will find described in these pages the many applications of ceramics in their lives. Additionally, they will be given the opportunity to understand how and why ceramics work in the applications described. For example, the author summarizes the historical evolution of high-temperature inorganic nonmetallic chemistry in the chapter "From Pottery to the Space Shuttle." He discusses how ceramics have formed the core of art since antiquity in "The Beauty of Ceramics," and how variations in the atoms of a single ceramic compound can change the mechanical and optical properties of the material in "Ceramics and Light." Ceramics are central to developments in bioengineering and medicine, energy and pollution control, and could revolutionize electronics through new nanotechnology research.

Our world revolves around ceramics on a daily basis; we may utilize a computer dependent upon a ceramic integrated circuit, gaze out glass windows, drive our automobiles powered with ceramic component engines, walk on concrete walks, eat from china dishes, admire a new glass sculpture, hit a few golf balls with a composite five iron, send data over high-speed glass fiber optics, or brush our teeth over a porcelain sink. The reader will gain a better appreciation of all of these applications in *The Magic of Ceramics*. The author has translated a very complex and technical subject with the inherent fundamentals of chemistry, physics, and mathematics into a readable, engaging, and interesting text.

It is also my hope that the readers will gain a better appreciation for the researchers, engineers, and technologists who dedicate their lives to better understanding the composition and properties of ceramic materials and to the development of new materials—even to the extent of manipulating individual atoms.

My rock-collecting days were ended when my mother inadvertently pulled open the top drawer of my dresser a bit too far and it fell rapidly to the floor, narrowly missing her feet. All those years of collecting had yielded a sizeable poundage of rocks. I eventually attended the University of Washington, where I was introduced to ceramic engineering by then department chair Dr. James I. Mueller. The department also had a NASA grant to help develop the ceramic tiles that cover the exterior of the Space Shuttle. It was enough to lead me through two degrees in ceramic engineering. A decade later, I was selected to be a Space Shuttle astronaut. My career as an astronaut continues and, as I now well know, the Space Shuttle program depends upon many ceramic material applications: from the quartz windows, to the computer components, to the heat resistance properties of the ceramic tiles. The reader will also learn much more than I knew during my rock-collecting youth. It is my privilege to have been a part of the ceramic engineering discipline and to provide this foreword for *The Magic of Ceramics*.

BONNIE J. DUNBAR, PH.D.
NASA Astronaut

Ceramics are amazing materials! *Some are delicate and fragile; others are so strong and durable that they are used to reinforce metals and plastics. Some ceramics are transparent. Others are magnetic. Many ceramics withstand temperatures many times the temperature of your oven and are untouched by erosion and corrosion that destroy metals in days. Ceramics have so many different characteristics and make so many things possible in our modern society that they seem magical.*

Without ceramics we wouldn't have television, miniature computers, the Space Shuttle, CDs, synthetic gemstones, or even cars. We wouldn't be able to refine metals from ores or cast them into useful shapes. We wouldn't have many of the modern tools of medicine such as ultrasonic imaging, CT scans, and dental reconstructions. How can ceramics do so many things? Seems like magic, doesn't it?

Have you ever seen a magician perform an amazing feat and wonder how it was possible? No matter how spectacular the illusion, there is always an explanation or trick, and often the trick is as fascinating as the illusion. The magic of ceramics is much the same—the feats and explanations are equally amazing. Reading this book will show you some of the magic that ceramics do and will explain the fascinating science that makes the magic work.

You'll learn how ceramics interact with light to produce great artistic beauty and to make the laser possible; how some ceramics can be stronger than steel and are used for inline skates and bullet-proof armor; how magnetic ceramics made the first computers possible and are the secret behind recording tape and CDs; and how a whole new field of "bioceramics" has emerged to enable miraculous medical cures and repairs. Ceramics touch and enrich our lives in so many ways! I take great pleasure in sharing some of that magic with you!

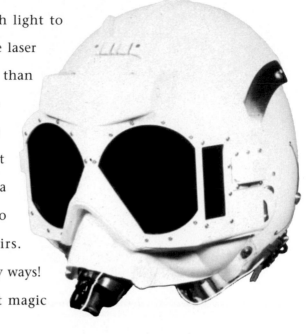

DID YOU KNOW?

- Some ceramics are so strong that a one-inch diameter cable could lift 50 automobiles

- Enough fiber-optic cable has been installed to go to and from the moon 160 times

- More than three million spark plugs are manufactured each day

- Ceramic automotive emission control systems have saved us from 1.5 billion tons of pollution since 1975

- Each year about one ton of cement is poured for each person on earth

- Some ceramics conduct electricity better than metals

- Diamonds, rubies, and cubic zirconia are ceramics

- Glass microspheres smaller than a hair provide a promising new liver cancer treatment

- Enough ceramic tiles are produced each year to pave a path 300 feet wide around the world

- Ceramic fiberglass house insulation has conserved more than 25,000,000,000,000,000 Btu of heat since 1938

The★ Magic of Ceramics

Our Constant Companions

CHAPTER 1

*I*t's hard to imagine the tremendous role that ceramics play in our everyday lives. Ceramics come in nearly infinite forms and behave in equally diverse ways. Nearly everything we do brings us in contact with either ceramics or something that was made using ceramics. In fact, ceramics are virtually our constant companions; they affect our daily lives in ways that border on the magical. If you think of ceramics only as decorative materials or "the stuff that dishes and toilets are made of," you're overlooking an important part of your world.

FROM STONEWARE TO SUPERCONDUCTORS

What, exactly, are these remarkable materials that have such an effect on our lives? One highly regarded professor and author (W. David Kingery, in his classic text *Introduction to Ceramics)* defines ceramics as "the art and science of making and using solid articles which have as their essential component, and are composed in large part of, inorganic, nonmetallic materials." Simply stated, most solid materials that aren't metal, plastic, or derived from plants or animals are ceramics.

As you might imagine from this definition, the term *ceramics* covers much ground: from *traditional ceramics* such as pottery, tile, and glass that date from antiquity to amazing new *advanced ceramics* that sport strange names like silicon nitride, aluminum oxide, and cordierite. Even synthetic gemstones such as ruby, sapphire, and cubic zirconia are ceramics. What would we do without glass or bricks or concrete? Although these traditional ceramics have been used for centuries, they are still a vital part of our lives. They're everywhere we look. Even advanced ceramics have entwined themselves in our daily lives in an incredible number of hidden, and often magical, ways. To initiate your entry into the world of ceramics, let's take a ceramic tour of your own everyday world. You may find it surprisingly familiar. I hope you'll then join me for a more in-depth look into the magic of ceramics.

CERAMICS IN YOUR HOME

Wake-up Call

Good morning! If your world's anything like mine, ceramics just woke you from a peaceful sleep. Chances are, your clock or clock radio has an alarm buzzer made from an advanced *piezoelectric* ceramic. This unusual ceramic vibrates with a loud noise when electricity is applied. If your clock has a quartz mechanism, as most clocks and watches now do, a tiny slice of vibrating piezo-electric quartz ceramic is the timekeeper. However, in this case the vibrations are so rapid and small that you can't hear them. Most people have never heard of piezoelectric ceramics, but piezoelectrics are the secret behind a wide variety of products ranging from underwater sonar (submarine searchers) to medical ultrasonic scans to "smart" skis. The secrets of these and other surprising piezoelectric applications are presented in Chapter 7.

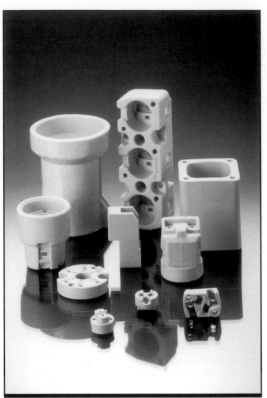

Figure 1-1. *Ceramic electrical insulation such as that used in houses and buildings.* (Photograph courtesy of Rauschert Industries, Inc., Madisonville, TN.)

Does the face of your alarm clock glow? This glow also may be due to a ceramic, one with the exotic name *electroluminescent*. The glow is caused by electricity passing through a very thin film of this special ceramic. The clockmaker can control even the color of the glow by selecting the presence of certain atoms in the ceramic. Electroluminescent ceramics also light the instrument panel in your car, the face of many wristwatches, and your cellular phone. Electroluminescence and other ways that ceramics interact with light or produce light are described in Chapter 4.

Perhaps you prefer to awaken to the sound of soft music from your clock radio. The radio itself is full of ceramic electrical devices (capacitors, insulators, resistors), all working together to soothe you or get you on your feet.

Before you jump out of bed, though, just lie there a minute, stretch your muscles, and survey your room to begin our tour. Note the glass mirror on the dresser and, possibly, the porcelain drawer knobs. Glass also may shield numerous pictures on your walls. The early morning sun streams in through your glass windows, and decorative glass covers your overhead lighting. The light bulbs in your nightstand lamps are glass, as are the vase on the table in the corner and the bottles of colognes, cosmetics, and lotions on your vanity. Although you can't see inside your bedroom walls, the studs are covered with ceramic plasterboard, and the electrical outlets and lights have hidden ceramic insulators. Some of the wall paint even contains ceramic pigments. Flip on the bedroom TV, if you have one, to catch the morning news. Your television is loaded with ceramic parts including the glass screen, the ceramic *phosphors* that produce the color, and numerous capacitors, insulators, resistors, and integrated circuits (we'll discuss these all in later chapters).

Well, it's time to drag yourself to the shower to freshen up. What do you see in the bathroom? More ceramics. As a matter of fact, your bathroom probably has more pounds of ceramic per square foot than any other room in your house. The mirrors and lights are glass. The toilet and sink are ceramic, and the tub is lined with porcelain enamel or constructed completely from glass-fiber-reinforced plastics. The floor, the counter around the sink, and the whole shower/tub compartment may be protected by colorful ceramic tiles. Besides offering beauty, these tiles provide an easy-to-clean surface that has dramatically improved sanitation and health. Ceramic tile has become so universally accepted that enough ceramic tiles are manufactured in the world each year to pave a road the width of three football fields encircling the entire Earth.

Like the bedroom, the bathroom also contains hidden ceramics. The water faucet valve that mixes hot water with cold and also seals against leaks is probably ceramic. The thermal and electrical insulation, and maybe even the heating element, in your hair dryer and in the space heater for cold mornings are ceramic. If you have an ultrasonic denture cleaner, the source of vibration is another piezoelectric ceramic. Even your electric toothbrush might contain ceramics doing their hidden magic. Surprisingly, ceramic powders may be hiding in some of your cosmetics, too. For example, *boron nitride* is commonly added to facial makeup. Boron nitride, which is white like talcum powder (also ceramic), is made up of tiny flat particles that smear onto a surface (such as your face) to produce a smooth, soft texture.

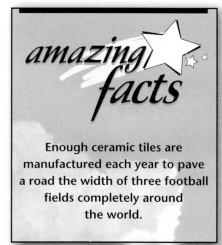

amazing facts

Enough ceramic tiles are manufactured each year to pave a road the width of three football fields completely around the world.

After showering and shaving, you may return to the bedroom to dress. The buttons on your shirt or pants could be ceramic. If so, they can withstand the most active person and the most abusive laundry situations. The stones in your rings may be cubic zirconia or even synthetic ruby or diamond. Do you wear eye glasses and a wristwatch? Your watch, like your clock, most likely has a glass cover, an electroluminescent dial, a quartz piezoelectric mechanism, a piezoelectric alarm, and even a lithium battery containing ceramics. If it's a digital watch, it has a liquid crystal display (not ceramic) sandwiched between special protective glass layers. Your watch is a marvel of miniaturization and precision. You may be lucky enough to have a special watch such as those made by the Rado company, with a cover glass fabricated from synthetic sapphire and its bezel and wristband links shaped from a special new ceramic called *transformation-toughened zirconia* (another ceramic that we'll discuss later). Such watches are nearly scratch-proof and works of art.

As you leave the bedroom and enter the hall, you may pass your children's rooms and perhaps another bathroom. Unless you're braver than I or have to venture in to wake the kids, you may want to use your already broadening knowledge to imagine the ceramics buried among their backpacks, shoes, and days-old laundry. In the hall, you should pass a smoke detector and perhaps a carbon monoxide detector. The indicator lights on both life-saving devices are made of light-emitting ceramic parts, and the sensor in the carbon monoxide detector is ceramic.

Figure 1-3. *Ceramic buttons manufactured by Coors Ceramics, Golden, Co. (Photograph by D. Richerson.)*

Figure 1-4. *Beautifully crafted Rado watch fabricated mostly from ceramics.*
(Photograph courtesy of the Rado Watch Co., Ltd., Lengnau, Switzerland.)

Figure 1-5. *Steps in the fabrication of the watch glass and bezel of a Rado watch.* *(Samples courtesy of the Rado Watch Co., Ltd., photograph by D. Richerson.)*

Food for Thought

Time for breakfast, at last! Your kitchen is positively loaded with ceramics: drinking glasses, dishes, storage containers, and cookware. Some of the dishes represent a technology unheard of only years ago. For example, you may have breakage-resistant Corelle dishes. These magical, bouncing dishes are created by heating special layers of ceramic that each respond to heat in different ways. When the newly created dishes are cooled, their inside layers shrink more than their outside layers. This shrinkage pulls the outer layers tightly together, making the ceramic dishes many times stronger than traditional dishes. Such plates rarely break when dropped or accidentally banged together because technology has turned them into superceramics of everyday life.

What other items do you see in your kitchen? Do you have a ceramic tea kettle on the stove, a slow cooker, or an electric mixer with glass bowls? How about the bottles in your spice rack, cupboards, or refrigerator? Open your refrigerator door. What holds the door closed? The seal is probably a thin strip of a rubberlike polymer filled with ceramic magnetic particles. You most likely also have a few ceramic magnets on the outside of your refrigerator door holding up messages, cartoons, or bits of family philosphy. (Ours are encased in colorful, plastic fuzzy creatures that hold up a cartoon of a mother with her hands in the air and a distressed expression, accompanied by the words, "Insanity is hereditary, you get it from your children.")

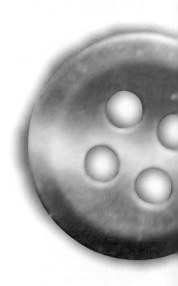

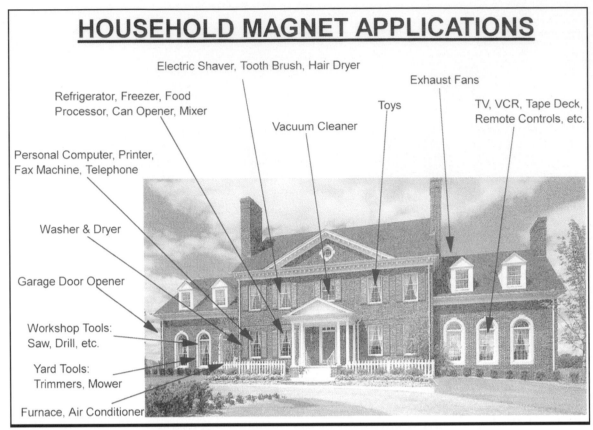

HOUSEHOLD MAGNET APPLICATIONS

Electric Shaver, Tooth Brush, Hair Dryer

Exhaust Fans

Refrigerator, Freezer, Food Processor, Can Opener, Mixer

Toys

TV, VCR, Tape Deck, Remote Controls, etc.

Vacuum Cleaner

Personal Computer, Printer, Fax Machine, Telephone

Washer & Dryer

Garage Door Opener

Workshop Tools: Saw, Drill, etc.

Yard Tools: Trimmers, Mower

Furnace, Air Conditioner

Figure 1-6. Ceramic magnets in a typical house. *(Illustration courtesy of Group Arnold, Marengo, IL.)*

Figure 1-7. (Below) Ceramic scissors and large kitchen knife by Kyocera. *(Photograph by D. Richerson.)*

Your kitchen probably also has a clock, a window, and maybe even a skylight, all containing transparent glass. We have a window over our sink. Hanging in front of the window are plants in ceramic pots cradled in macrame slings with ceramic beads. An artistic, circular stained-glass image of a butterfly on a flower enhances the early morning light.

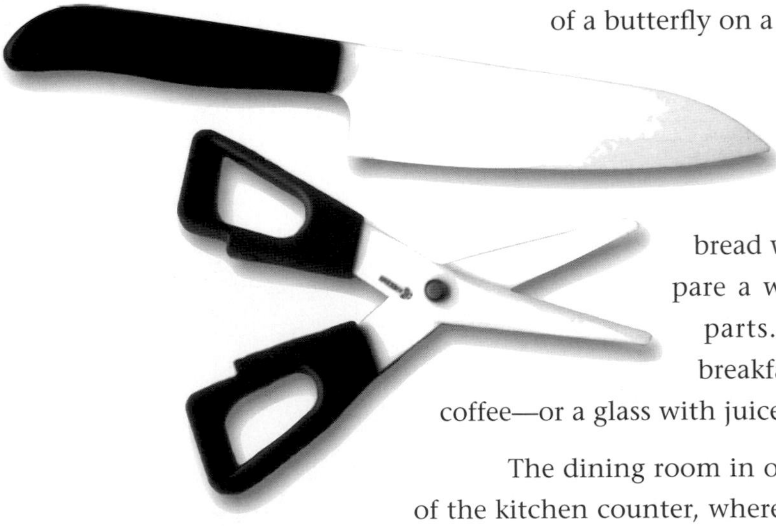

After all this work looking for ceramics, you're probably hungry. You may cook your bacon and eggs in a ceramic skillet on a ceramic stovetop, cut your bread with a ceramic knife that never dulls, and pre-pare a waffle or toast in an appliance with ceramic parts. And you almost certainly will serve your breakfast on a ceramic plate, fill a ceramic mug with coffee—or a glass with juice or milk—and head for the dining area.

The dining room in our house is a small area just on the other side of the kitchen counter, where a large, glass-topped table consumes most of the space. Somehow we managed to squeeze in chairs, a bookcase, a TV stand complete with TV, a series of shelves under the window completely covered with plants (many in ceramic pots), a small refrigerator that mysteriously consumes at least a six-pack of soft drinks each day, and a large buffet. The top of the buffet is covered with family photographs in glass-faced frames. Each holiday, the top of the buffet sprouts a new assortment of seasonal ceramic knick-knacks: a ceramic Santa Claus collection, an Easter egg tree with ceramic and real eggs, and so on throughout the year.

The Rest of the House

The family room in most homes is another source of ceramics. Think a bit while you eat breakfast. Does your family room have a television? You might have a stereo, too, which contains numerous ceramic components. Family rooms usually have lots of pictures, lamps, and candy dishes. You might have a chess set with ceramic pieces, a game of Chinese checkers with glass marbles, or a cribbage

Figure 1-8. *The author's dining room, showing the table set with ceramic dishes, and glasses and the buffet in the background, decorated with ceramic knickknacks. (Photograph by D. Richerson.)*

board with ceramic markers. Some family rooms have pianos, which often are topped with ceramic knickknacks or family pictures in ceramic frames. What is the source of lighting in your family room? Do you have a skylight or a patio door or windows? Our family room is well lit with the soft luminescence from fluorescent lights, which produce the light with a ceramic phosphor as we'll discuss in Chapter 4.

The living room or great room probably has the least ceramics in your home. Although it has no appliances, it may hold an assortment of

Figure 1-9. *Hummel ceramic figurines. (Photograph by D. Richerson.)*

glass-covered art, ceramic figures, and potted plants. The end tables and coffee table may be glass topped. Our great room has a Tiffany-style light fixture over the formal dining table. If you have a fireplace in your living room, it undoubtedly has a ceramic brick hearth.

Most homes have basements, or at least crawl spaces. Our basement has an office, a laundry room/furnace room, a recreation room, a hobby room, and lots of space to store the off-season ceramic knickknacks.

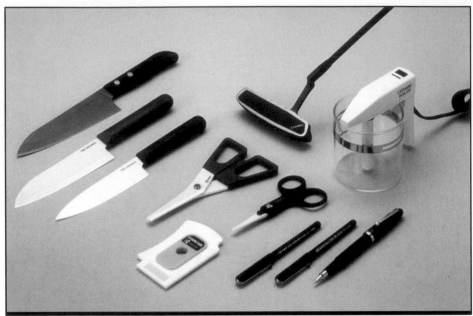

Figure 1-10. *A variety of ceramic items, including the tips of ball-point pens, scissors, knives, and a golf putter. (Photograph courtesy of the Kyocera.)*

The computer, fax, and printer in my office contain numerous ceramic parts and would not work without them. If you have a home office, your desk is likely filled with small objects that may boast ceramic parts: ceramic-tipped letter openers, scissors, and even pens with ceramic nibs. Do you have an office coffee pot? How about a touch-tone telephone? A clock? A tile floor? In the laundry room, your washing machine has a ceramic seal to keep the water from flooding the room, and your gas dryer has a ceramic igniter to light the gas each time you turn on the dryer.

Off to Battle

Fiberglass housing insulation was introduced in 1939 and has resulted in estimated energy savings of more than 25,000,000,000,000,000 (25 quadrillion) Btu.

Well, the light coming in higher through the windows should tell you it's time to get going. What would life be like without glass to let the light in and keep the weather and bugs out? In fact, ceramics do duty outside the house as well as inside. Your house, as mine, may be covered with brick. The walls and attic are probably filled with fiberglass (another hidden ceramic) insulation to protect you from the heat and cold. Fiberglass housing insulation, introduced in 1939, has resulted in estimated energy savings of more than 25,000,000,000,000,000 (that's 25 quadrillion) Btu, enough to provide all the energy needs of North America for two years. Even the shingles on the roof contain glass fibers. You'd certainly notice if these ceramic materials weren't there, especially if it's rainy or cold this morning. Behind our house, a concrete patio supports a glass-fiber-reinforced spa with ceramic working parts and pump seals.

For most of us, the trip to work or school starts in the garage. Our garage has a concrete floor, brick walls, fluorescent lighting, and an automatic garage-door opener containing ceramic parts to make it open and close on command. The driveway, sidewalk, and porch also are concrete. Concrete, in fact, may be the most abundant ceramic in the modern world. Each year, roughly one ton of concrete is produced for every man, woman, and child on Earth, now about 6,000,000,000 of us in all.

About one ton of concrete is produced each year for every man, woman, and child on Earth!

Most of us use some type of motor vehicle to get to work or school. Cars, trucks, and buses are loaded with ceramics. As a matter of fact, the typical automobile has more than 100 critical ceramic parts. (We'll talk about many of them in Chapter 9.) As you drive to work or school, you use signal lights enclosed in glass to safely navigate through intersections. If you drive at night, your safety is further protected by the yellowish sodium vapor lamps lining urban streets. These lamps owe their existence to advanced ceramics, as you will learn in Chapter 12. The buildings you pass are built mostly of brick, concrete, and glass. You may even drive part of the way on a concrete highway. The painted lines on the road contain reflective ceramic particles that make them easy to see in the dark and also more resistant to wear and tear. Some of the traffic signs contain ceramic powders to make their warnings more visible in dim light. The cellular phone you use to keep in touch with the office or your family uses ceramic filters to screen out radio waves, TV signals, and police communications that would interfere with proper reception.

CERAMICS ON THE JOB

The building where you work or go to school, like those you passed on your way, probably is constructed with ceramic materials similar to those in your home. Most office buildings house computers, fax machines, copiers, printers, calculators, and touch-tone telephones, all of which contain critical ceramic parts. Schools have much the same equipment. One of your co-workers or one of your children's teachers may wear a pacemaker to control his or her heartbeat. Pacemakers have ceramic sheaths that carry wiring to them from their power pack, ceramic capacitors that enable them to store their life-giving energy, and often ceramic cases to protect them from body chemicals. Another teacher or co-worker may have a ceramic hip replacement. Most likely, you and most of the people you

Figure 1-11. *Ceramic labware for chemical studies and analysis, manufactured by Coors Ceramics, Inc., Golden, CO.* *(Photograph by D. Richerson.)*

know and work with have experienced some sort of dental or medical work involving ceramics within the past few years.

Your work day may involve ceramics in more direct ways, too. Obviously, mine does, but many other workplaces, such as school cafeterias and science laboratories, machine shops, medical facilities, restaurants, city water plants, and even grocery stores, use ceramics in important ways every day. Some workplaces have laboratories with ceramic equipment for chemical analysis or experimentation.

CERAMICS ELSEWHERE

If you have to stop at the supermarket on your way home from work or after you pick the children up from school, you may be surprised to learn that ceramics create everyday magic there, too. The laser scanner at the checkout counter is an example. The laser itself is made possible by ceramics. When laser scanners first came out, the transparent "window" through which an item's bar code is read was made of glass. Glass, however, tended to get scratched by all the cans and other items sliding over its surface. (You've probably tapped your foot in a checkout line while the cashier tried several times, without success, to get the scanner to read the

Figure 1-12. *Transparent sapphire scanner plate for a laser bar code reader in a supermarket.* *(Photograph courtesy of Saphikon Inc., Milford, NH.)*

price on your item.) For more efficient operation, scanner plates now are fabricated from sheets of synthetic ruby or sapphire, which are much harder than glass and resist scratching.

Do you pass any electrical power transmission lines on your way home from work? The odd-shaped insulators that separate the electricity-carrying metal wires carrying the electricity from each other and from the tall metal or wood towers are ceramic. And ceramics are required in many ways during the generation of electricity, whether the plant is powered by coal, nuclear energy, hydroelectric power, gas turbines, or even wind. The importance of ceramics now, and this increasing importance in the future, to energy production and pollution reduction are described in Chapter 12.

You may pass a chemicals plant, a paper mill, a semiconductor plant, or an automobile manufacturing plant on your drive home, all of which use ceramics. It's hard to imagine an industry that does not depend heavily on ceramics. Some fascinating examples are included in Chapters 10 and 11.

Is it time to see your doctor for a yearly checkup? Ceramics do heavy-duty work in doctors' offices, clinics, and hospitals as key components in X-ray, CAT-scan, laser, and ultrasonic-imaging equipment. They are sterile containers for blood, urine, and bacterial cultures. Ceramics even are used extensively in surgery, especially to allow modern noninvasive endoscopic examinations and surgical procedures such as knee repairs and gall bladder operations. You'll be amazed at the medical applications of ceramics reviewed in Chapter 8.

Do you go out to eat with your family at the end of a long day? Ceramics are busy at work in restaurants just as they are in your kitchen at home. Even fast-food restaurants and the food services in malls and gasoline stations depend on ceramics. For example, next time you fill your glass from a soft-drink dispenser, take a closer look. The mixing valve that meters the syrup and carbonated water and guides the flow into your glass is often ceramic.

CERAMICS AT PLAY

Of course, most of us have another important life after work and on weekends. We may go to the lake, catch a game of tennis, drop the kids off at a soccer match, or take in a baseball game with our family. Most sports rely on ceramics in one way or another. Baseball bats sometimes contain a ceramic core or ceramic fibers to strengthen them. Skis (for both water and snow), paddles for canoes and kayaks, hockey sticks, and tennis rackets are all strengthened by ceramic fibers. Nearly all recreational boats are reinforced

Figure 1-13. *Golf putter and cleats made from ceramic materials by Coors Ceramics, Inc., Golden, CO.* (Photograph by D. Richerson.)

with glass fibers to protect them from damage as they streak through the waves. Bowling balls often have ceramic cores that increase the amount of energy transferred from the ball to the pins by about 5%. Ceramic core bowling balls "hit harder." Golf-club shafts frequently are reinforced with carbon fibers to increase strength and stiffness, and the heads of some putters are made of specially toughened zirconia (an advanced ceramic that will be explained in Chapter 5), as are new golf cleats that provide good traction without damaging the greens. Maybe your favorite entertainment is travel, or watching television, or astronomy. Guess what? You're right—ceramics are there to help you again!

OVERVIEW: MOVING ON

It's time to move on, even though we've barely scratched the surface of the world of ceramics. Ceramics truly are everywhere. Their amazing range of properties and uses has brought us many of the remarkable technologies and products that define modern civilization. (That's where piezoelectrics, phosphors, fluorescence, zirconia, and electroluminescence come in.) Ceramics also add to the quality of our life through their beauty. They enhance our lives by their constant usefulness. Equally important, they play a critical role in electronics, communications, transportation, manufacturing, energy generation, pollution control, medicine, defense, and even space exploration. They have filled our past, and they enrich our present. They will pave the path to our future. Before you learn some of the science behind the magic of ceramics and explore important uses for ceramics, let's go back in time and see how ceramics evolved from its origin as pottery to high-technology modern applications such as in the Space Shuttle.

From Pottery to the Space Shuttle

*H*istory and ceramics are intertwined. Advances in civilization have always followed advances or innovations in materials. As archaeologists and anthropologists tell us, one of the first steps in human development was taken when early cultures learned to use natural materials, such as wood and rock, as tools and weapons. The next step began when they learned to use rocks to chip other rocks, such as chert and obsidian (volcanic glass), into more efficient tools and weapons. So important was this use of natural ceramic materials that the prehistoric time in which it occurred is now referred to as the Stone Age. But the Stone Age was just the beginning of the use of materials to improve our standard of life. Eventually, people learned to make pottery; to extract and use metals (the Bronze Age and Iron Age); to produce glass; and to make bricks, tiles, and cement. Much later materials made possible the Industrial Revolution, the harnessing of electricity, and the "horseless carriage." Only within the last two generations, during the time of our parents and grandparents, have materials ushered in the Age of Electronics, the jet airplane, near-instantaneous worldwide communications, and the exploration of outer space.

PRODUCTS AND USES

Traditional Ceramics
Earthenware pottery
Stoneware, porcelain
Bricks and tiles
Mortar, cement, plaster
Glass containers
Furnace linings

Modern Ceramics
Spark plugs
Tempered glass
Synthetic gemstones
Space shuttle tiles
Quartz watch
Cellular phone
Miniature electronics
Lasers
Medical devices
Fiber optics

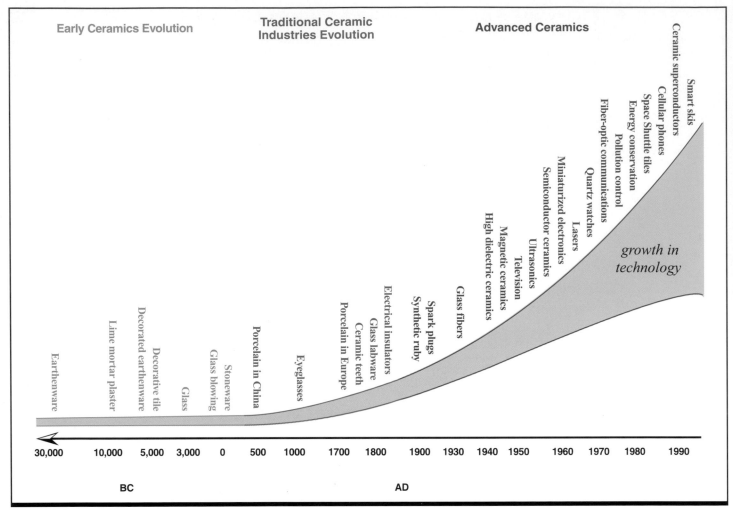

Figure 2-1. *The historical evolution of ceramics.*

Ceramics are important partners with metals and plastics in our modern civilization. Figure 2-1 highlights some of the key ceramics developments throughout history. The evolution of ceramics was slow for many centuries but has virtually exploded in the 20th century. In this chapter, we take a journey back in time to see how ceramics have evolved from the earliest dried clay articles to our amazing—often magical—modern ceramics.

EARLY CERAMICS

The Birth of Pottery

Let's imagine what life was like in the Stone Age 60,000 years ago. There were no stores, houses, or cities and not even a written language. Families used caves for shelter and wandered from place to place hunting food. They had no concept of a metal or a plastic. The only materials they knew were the natural materials surrounding them: rocks, plants, and the hides and bones of animals. Rocks are Mother Nature's ceramics. Primitive people shaped axes and other tools by beating these natural ceramics together. They learned to chip arrowheads and spear points. Because there was no written language, the stone artifacts and other items buried in the floors of ancient cave dwellings are the only evidence archaeologists and anthropologists have to study these early humans.

As the centuries passed, cave dwellers began to draw pictures on the cave walls using colored soil mixed with water. They discovered that some types of soil (which we now call *clay*) became pliable when wet and could be molded into shapes, such as the bison found in Tuc d'Audobert Cave in France. They observed that the clay became rigid when dried and hard like stone when placed in the fire. This discovery represented the birth of pottery, the first true man-made ceramics. We now call this simple pottery *earthenware*. In spite of the ancient origins, earthenware is still made today in nearly every corner of our world.

We don't know exactly when our ancestors learned to mold and fire ceramics, but archaeologists have guessed around 30,000 years ago. The oldest archaeological site found so far is in the Czech Republic and dates to about 27,000 BC. This site had a fire pit that appeared to be designed specifically for firing pottery. Animal and fertility figurines were found in and around the pit.

Figure 2-2. *Early example of images molded in clay. Mystical clay sculptures of bison in Tuc d'Audobert Cave in France, around 14,000 years old.* (Courtesy of the photographer, Count Robert Bégouën.)

The Emergence of Civilization

Pottery was an important innovation that helped mankind make the transition from a nomadic lifestyle to one of stable settlements. People learned to make earthenware containers for cooking and food storage, and they became less dependent on following their food sources endlessly from place to place. They began to form settlements to which they could return after a long day's hunt and store their food until it ran out. No longer exclusively nomads, these people now had time to put seeds into the ground and wait for them to grow into edible plants and grains. Agriculture was born, and ceramics were there to store and protect the harvest.

Early earthenware pottery dates back in time to at least 30,000 years ago.

Pottery containers allowed travelers to wander farther and farther from home, taking food along with them for long journeys. Extended travel by boat became possible, permitting exploration and the spread of civilization as people began to trade the food, wares, and pottery with others from ever-more-distant lands. Written records were not left to tell us about these travels and trade routes, but archaeologists have been able to piece together some of the puzzle of history by studying shards of pottery, because each culture created its own distinctive pottery. For example, each culture evolved its own style and decorations for earthenware. Some cultures made pots with one color of clay and painted designs or images on the surfaces after firing, using a different color of clay mixed with water, much like the earlier cave paintings. The porous surface of this earthenware allowed the clay paint to soak in a little, so that the decorations didn't rub off during handling. Other potters scratched (incised) patterns or images into the surfaces of their pots.

Ceramics in the Age of Metals

Although early ceramic materials were a life-changing innovation, they were still very fragile and broke easily. By about 4000 BC, people had learned how to separate metals from the natural minerals, or *ores,* in which they occurred inside the earth. These new materials were much tougher and stronger than ceramics, and they could be beaten into useful shapes with a hammer or other tool or melted and poured into a shaped mold. The Stone Age gave way to the Bronze Age, which later was displaced by the Iron Age.

The "metals ages" had a dramatic impact on civilization but did not replace ceramics. In fact, ceramics became even more valuable because of the very quality that had been discovered by stone-age people so many years earlier: resistance to extreme heat. Extracting metals from ores required high temperature, and ceramics were the only materials that could withstand such temperatures. Even after the metals had been extracted from their ores, ceramic containers (known as crucibles) were required for melting the metals. The metals then were poured into ceramic molds of various shapes and cooled to form tools and other objects, both beautiful and useful.

Pottery continued to evolve during the Bronze Age. One key to early Bronze Age ceramics innovation was improvement in the design of the chamber (*kiln*) in which potters fired their ware, in order to reach higher temperatures. Higher temperature firing reduced porosity and increased the strength of the ceramics, so the potters could make thinner-walled pots with flared bottoms and curled-over rims mimicking bronze. Potters also invented a new type of kiln that had two chambers, one for the fuel and one for the ceramic ware. This major breakthrough paved the way to new and exciting modes of decoration. Because the fire didn't directly touch the earthenware anymore, flame-sensitive ceramic paints could be applied to the pot before it was put in the kiln and then fired on to become a permanent part of the pot.

An even more exciting innovation was the creation of glazes. A potter's glaze is a glassy coating that not only can seal the surface of the porous earthenware against leakage of liquids, but also makes possible an endless variety of decorations. Early glazes probably were discovered around 3500 BC by potters trying to imitate the precious blue stone lapis lazuli. Small beads were carved from soapstone (talc) and coated with a powder of ground-up azurite or malachite (natural ores of copper with blue and green color). When fired, the coating interacted with the soapstone to yield a thin layer of colored glass. The potters probably borrowed this idea and started experimenting with different combinations of crushed and ground rock mixed with water and painted onto the surface of pots. They discovered mixtures that worked, that completely coated the surface of their earthenware with a watertight glassy layer. As the centuries passed, potters learned to produce glazes in many colors and textures and even in multiple layers, by using multiple firings at different temperatures. We will discuss some of these techniques and creations in the next chapter, "The Beauty of Ceramics."

amazing *facts*

Glazes are glassy coatings that help make earthenware containers watertight. Early forms of colored decorative glazes date back to around 3500 BC.

EVOLUTION OF TRADITIONAL CERAMICS

The Invention of the Potter's Wheel

Metals were expensive and could only be afforded by the wealthy. Much more affordable to the average person, pottery ultimately became an important part of every household and was the first *traditional ceramic*. An important innovation that helped pottery become affordable was the potter's wheel, which was invented around 2000 BC in both Mesopotamia (east of the Mediterranean Sea and south of the Caspian Sea in the valleys of the Tigris and Euphrates Rivers) and Egypt (south of the Mediterranean Sea along the Nile River). The first potter's "wheel" was probably a mat on which a flat stone or broken piece of pottery could be slowly rotated by hand while the potter formed a mound of moist clay into a hollow, circular shape. Improved potter's wheels could be rotated by the potter's foot or by an assistant, so the potter could have both hands free to mold and shape the clay. The potter's wheel dramatically increased the number of pieces that could be produced per day and contributed to broad availability of earthenware to the average person.

amazing facts

The potter's wheel, invented around 2000 BC, revolutionized pottery making.

Earthenware spread throughout the ancient Western world and evolved independently in the Far East. During the Roman Empire (about 100 BC to AD 300), mass-production methods were established to make enough pottery to meet the needs of the Roman army and growing cities. About this same time, during the Han Dynasty (207 BC to AD 230), ceramics use blossomed in China to become an important part of daily life as wine vases, storage jars, cooking vessels, ladles, dishes and bowls, kettles, candlesticks, and even small tables.

New Types of Pottery

As mentioned earlier, potters learned that firing their ware at higher temperature resulted in a stronger, less porous ceramic pot. Chinese potters were especially intrigued by this technique and were much more aggressive than Western potters in experimenting with different kiln designs and recipes for ceramic raw materials. They succeeded in building kilns that could fire at around 2200°F (about 1200°C), nearly five times hotter than a kitchen oven. Pots fired at such high temperature had a low enough porosity that they could hold water with no leakage even without a glaze. We now refer to this type of ceramic as *stoneware*.

The Chinese slowly refined stoneware during the Shang Dynasty (1500 to 1066 BC) and Chou (Zhou) Dynasty (1155 to 255 BC).

A key discovery was the use of a white clay call kaolin, which needed a high temperature to fire properly. Pots made with kaolin were nearly white in color, rather than the various shades of brown and reddish tones of earthenware and prior stoneware.

By around AD 600, Chinese potters had discovered another secret ingredient that they called petuntse. This material was a natural rock in China that could be crushed into a fine powder and added to stoneware. When fired at a very high temperature (1300°C), the petuntse interacted with the kaolin and powdered sand in the recipe to form some glassy material similar to a glaze, but that was distributed throughout the ceramic rather than just at the surface. The resulting ceramic was pristine white and translucent (light could glow through the thin walls). A truly magical creation, this ceramic even could make a beautiful sound. When tapped with a fingernail or a hard object, the ceramic would ring with the crystal-clear tone of a fine chime or musical instrument. When Marco Polo returned to Europe in 1292 from his epic journey to the Orient, he described this magical ceramic as *alla Porcella*—"having the appearance of a delicate, shiny seashell"—because it reminded him of a delicate seashell called porcellana in Italian. We now refer to this fine ceramic as *porcelain*.

Europeans were completely captivated by the beauty of porcelain and tried for centuries, without success, to duplicate it. Finally, in about 1710, Ehrenfried Walther von Tschirnhausen and Johann Friedrich Böttger accomplished the feat at Meissen in Saxony, Germany. Meissen attracted the best ceramic designers in Europe and quickly became the source of exquisite porcelain vases, sculptures, figurines, and other ornate creations. By the mid 1700s, Josiah Wedgwood, in England, introduced an assembly-line approach to the fabrication of porcelain and similar fine ceramics (bone china), making them affordable to a broader cross section of society. Porcelain joined earthenware, bricks, and tiles as a mass-produced traditional ceramic.

> **amazing facts**
>
> Porcelain emerged in China circa AD 600; Europeans were not able to duplicate it until about 1710. The name *porcelain* was coined by Marco Polo in 1292.

Tiles, Bricks, and Cements

Through the centuries, earthenware, stoneware, and porcelain production have become industries passed on from generation to generation and are an important part of our traditional ceramics heritage. But other forms of ceramics also have evolved from ancient times and become important traditional ceramic industries: tiles, bricks, and cements. Decorative tiles, for example, date back to at least 4th millennium BC (4000 to 3000 BC) Egypt. By around 2000 BC, tile, both glazed and unglazed, had spread to Asia Minor and Mesopotamia, and became important, along with brick, as a material for construction and decoration.

Tiles were made individually, by hand, using the same raw materials that were used for pottery—mostly clay and sand. Designs were either cut into or pressed into the tile while the clay was still damp and slightly pliable. Surprisingly, no major innovations in decorative tilemaking occurred until the industrial revolution, about 130 years after the death of William Shakespeare. In the 1740s, an Englishman, Richard Prosser, invented transfer printing. This process was a new method for quickly forming a glazed design using a wood block tool and paper and then transferring the design onto the surface of the tile. Using transfer printing, two workers could produce as much tile per day (1200 tiles, roughly enough to cover a floor 35 feet long by 35 feet wide) as a whole factory of 100 people had been able to produce before.

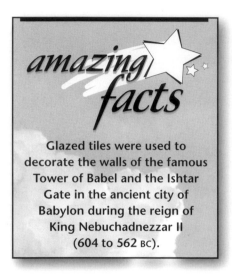

Glazed tiles were used to decorate the walls of the famous Tower of Babel and the Ishtar Gate in the ancient city of Babylon during the reign of King Nebuchadnezzar II (604 to 562 BC).

Another key invention 100 years later was dust pressing. Until then, tiles had been made by hand from wet, moldable clay and took a lot of time to prepare, shape, and dry. Dust pressing was a process that sandwiched nearly dry clay (plus other tile ingredients) between two metal plates and then forced the plates against the clay by the turning of a large screw. High enough pressure could be applied to compress and compact the powder into the desired tile shape. Since the tiles made in this way were nearly dry, they could be fired immediately after removal from the press, rather than going through a lengthy drying procedure. The cost of making tiles went way down, and more people—even the poor—could afford to use them in their homes. Tiles had clearly reached the status

Transfer printing, invented in the 1740s, enabled two people to produce as much decorative tile in one day as 100 workers could produce previously.

of a mass-produced traditional ceramic, but even more important, dust pressing became a key technology in the later development and commercialization of advanced ceramics.

Glass

One of the most important and magical categories of ceramic materials is glass. To understand how glass is different from other forms of ceramics (such as stoneware and tile), we need to picture in our minds the internal structure of glass. As shown in Figure 2-3, a typical ceramic is made up of many particles, or grains, tightly bonded together, often mixed with tiny open spaces called *pores*. Each grain is a crystal, just like the grains of quartz, mica, and feldspar that you see sparkling in a piece of granite rock. The grains in ceramics, however, are generally too small to be seen easily without a microscope. Just as you can see a line or boundary between the different grains in granite, each grain in a ceramic is separated by a boundary. Not surprisingly, this is called a grain boundary. The whole combination of grains, grain boundaries, and pores is referred to as the microstructure of the ceramic, and this type of ceramic is referred to as polycrystalline (made up of many tiny crystals). In contrast, a glass has no grains or grain boundaries. It has been fired to a high enough temperature to completely melt the grains into a uniform liquid, and then cooled so quickly that new grains do not have time to form. Glass is like a liquid frozen in time. Because glass is so uniform, and usually has few or no pores, it lets light pass through, providing us with a whole new appearance, different from that of polycrystalline ceramics.

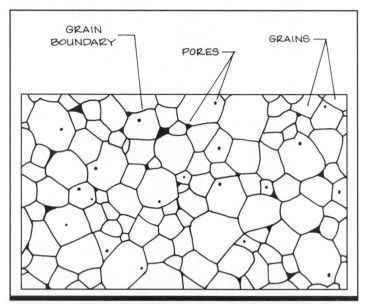

Figure 2-3. Features in the microstructure of a typical polycrystalline ceramic.

Figure 2-4. Cast-glass head of Amenhotep II. From the early 18th dynasty of Egypt, about 1436 to 1411, originally cast as blue glass, but weathered at the surface to a buff color; about 1.6 inches high (Courtesy of The Corning Museum of Glass, Corning, NY.)

Figure 2-5. Core-formed glass amphoriskos. Probably from the Eastern Mediterranean Sea region in the first century BC; and similar in style to the core-formed vessels from Egypt dating back to at least 1400 BC; about 9.4 inches high. (Courtesy of The Corning Museum of Glass, Corning, NY.)

The glaze development we discussed earlier, which dated back to about 3500 BC, was the first step in making objects from glass. By around 2500 BC, Egyptian potters had developed a special type of glaze we call faience. They ground up common sand into a powder and mixed it with wood ashes and natron. Natron is a material, found in some dried lake beds, that contains the chemical element sodium. Wood ashes contain the chemical element potassium. Although the ancient potters had never heard of potassium or sodium, they learned from trial and error that these materials acted as fluxes to reduce the melting temperature of the sand. When such materials were painted onto the surface of a pot and fired, a smooth glassy coating resulted.

Sometime between 2000 and 1000 BC, potters learned that they could melt faience-type mixtures in an earthenware container and pour this molten glass into another shaped clay container (a mold). After cooling, the clay could be removed to reveal the *cast glass* object.

Early glassworkers also learned to make hollow glass containers that were quite beautiful. The workers molded a core from a mixture of dung and wet clay into the shape they desired for the inside of their hollow container. After drying the core, they wrapped hot strands of semimolten glass around the core. They probably did this by sticking a piece of ceramic into the molten glass and pulling out a flexible strand of hot glass, similarly to the way candymakers pull taffy. Working quickly, they were able to wind different, colored strands of glass around the core before the glass cooled enough to become rigid. The glassworkers then reheated the glass-wrapped core until the glass was again soft and used a tool (probably ceramic) to press the glass strands together to form a continuous, smooth surface. After the ceramic cooled, the clay/dung core was scraped out to leave a hollow glass container.

Glassblowing was invented during the time of the Roman Empire, about AD 50..

Glass was precious to the early Egyptians and was only owned by pharaohs, high priests, and aristocrats. Glass did not become available to the general populace until around AD 50, during the time of Julius Caesar, when the art of glassblowing was invented somewhere along the Syrian-Palestinian coast. Glassworkers learned to attach a blob of molten glass to the end of a long hollow tube and blow gently through the tube to produce a thin bubble of glass. Hollow glass containers could now be formed quickly and efficiently, and glass became available to nearly everybody. Through the centuries, glassmaking became another important traditional ceramics industry for the production of containers, tableware, eyeglasses, and eventually windows.

MODERN CERAMICS

Earthenware, stoneware, porcelain, bricks, tiles, glass, and cements all have become traditional ceramic materials that are still vital today in virtually every part of the world. They are still made from natural materials dug out of the ground. But important changes began to unfold in the 1800s that have led to whole new classes of ceramics made using specially purified, or even synthetic, starting materials and that accomplish things traditional ceramics never could do. These new ceramics, referred to as *modern ceramics, fine ceramics,* or *advanced ceramics,* paved the way to our modern civilization.

Piecing the Puzzle Together

The history of modern ceramics is a bit like a puzzle that involves fitting many pieces together to reveal the whole picture. The discovery of electricity, early advances in chemistry, and the invention of the automobile all contributed to this ceramic puzzle. Even people's ancient search for the magic to create their own precious gems, such as rubies and diamonds, played a role in the development of modern ceramics.

Electricity. The need for special materials arose from discoveries about electricity. Remember the story of Benjamin Franklin's experiment in the late 1700s with a kite and lightning? At that time, scientists could see electricity, but they had no idea how to create or control it. The 1800s were a great century for electrical discovery and invention. Scientists learned not only how to generate and store electricity but also how to guide this mysterious force through metal wires to energize lights and other devices. Along the way, they also learned that ceramics, especially glass and porcelain, did not conduct electricity. This property made ceramics excellent *insulators* to keep electricity from going places it wasn't wanted. Existing traditional ceramic materials, however, couldn't keep up with the needs of a world steadily filling with electrical appliances and devices. It was time for new, improved ceramic materials: the first piece of the ceramic puzzle.

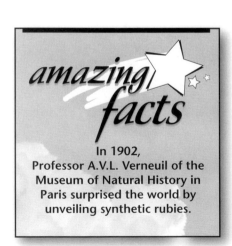

Chemistry. In the 1800s, scientists also began to understand some of the basic principles of chemistry. They learned that all matter is composed of tiny parts, called *atoms,* which in turn are made up of smaller parts called *electrons, protons,* and *neutrons.* The atoms are the basic building blocks of matter, the chemical elements. With their new knowledge, the scientists figured out how to identify exactly what elements were in a substance (its *chemical composition*) and how those elements were arranged physically to give the substance its unique shape and qualities (such as *microstructure*). Learning about chemical composition and microstructure helped the scientists to understand the relationship between how a substance acts and what it's made of. They discovered that impurities—extra elements that don't belong—in the materials from which ceramic objects were made could cause problems in the ceramics themselves. For example, iron oxide (which we all know as rust) in clay gave ceramics made from that clay a reddish or brownish color and also made the ceramics less effective as electrical insulators. On the other hand, white porcelain and glass, which didn't contain iron, worked much better as electrical insulators.

Ceramic engineers and chemists began to experiment with the chemical makeup of the raw materials from which ceramics were made. They learned how to refine these materials to remove impurities such as iron and also how to create, or *synthesize*, artificial new raw materials. These chemical advances were another important piece of the ceramic puzzle.

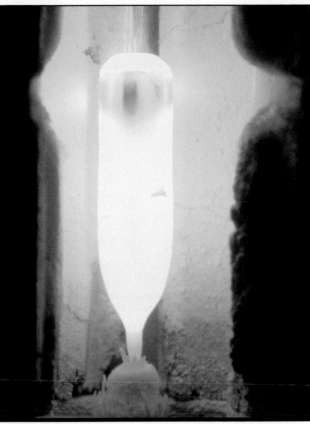

Figure 2-6. *Single-crystal sapphire boule being "grown" by the Verneuil process. The colorless, clear boule will be cut into slices and ground and polished into watch faces for Rado watches; sapphire is much harder and more durable than glass and incredibly resistant to scratching.* (Courtesy of the Rado Watch Co., Ltd., Lengnau, Switzerland.)

Precious Gems. Our age-old dream of creating precious gemstones, such as ruby and sapphire, added other pieces to the puzzle. Once scientists began to understand chemistry, they learned that ruby and sapphire are made from a special combination of aluminum and oxygen atoms that is described scientifically as *aluminum oxide* (Al_2O_3) and usually referred to by ceramists simply as *alumina*. Chemists found out that ruby is red because it contains a tiny amount of chromium along with its aluminum oxide. There's a bit of aluminum oxide in the Earth's crust (called *corundum*), but we rarely see it as the gem-quality crystals of ruby or sapphire that we so admire.

In the nineteenth century, researchers tried to change corundum into gems by heating it to a high temperature. They failed. Others tried chemical synthesis—building the gems from scratch. In 1837, the French scientist Marc Antoine Augustin Gaudin heated a chemical compound called *ammonium alum* (which turns into aluminum oxide when it's heated) with a high-temperature torch. He actually did produce some small crystals of corundum, in much the same way that sugar crystals form along a string left hanging in a glass of sugar water as the water evaporates; but the crystals were not gem quality, not good enough to be used in jewelry. Many alchemists, scientists, and entrepreneurs tried diligently throughout the 1800s to synthesize gems, but no one was successful. Then, in 1902, Professor A.V.L. Verneuil of the Museum of Natural History in Paris surprised the world by unveiling synthetic rubies.

Verneuil used chemistry to synthesize a pure powder of alumina with a little chromium oxide, the building blocks of ruby. He then trickled the powder carefully through the flame of a special torch that burned hydrogen and oxygen at an extremely high temperature—high enough to melt the ceramic powder. The melted alumina particles landed on a pedestal-shaped holder onto which Verneuil had placed a small piece of ruby called a *seed crystal*. Verneuil kept the flame very steady and slowly moved the pedestal away from the torch flame. Magically, a rounded, transparent single crystal, a *boule* (or ball) of ruby "grew" around the seed crystal on the top of the pedestal. Later, Verneuil created blue sapphire by adding another chemical, titanium, along with some iron, to the alumina powder. Since then, ceramists have obtained a whole range of other artificial gemstones in many different colors by adding various chemicals to an alumina starting powder.

Verneuil's rubies attracted international attention. Within a few years, artificial gemstones and other products, including "jewel" bearings for watches and scientific instruments, were common. Nearly 440,000 pounds (about the weight of 73 three-ton elephants piled one on top of another) of such stones were produced in the world every year before the invention of the quartz watch dramatically reduced the need for jewel bearings.

The process for making single-crystal alumina has changed quite a bit now. Much larger gems of higher quality than those grown using the Verneuil method are made by new techniques, such as the heat-exchanger process developed in the late 1960s. In this process, the alumina powder is put into a crucible made of the exotic metal iridium and melted above 2060°C in a high-temperature furnace. The bottom of the crucible is cooled to below the melting temperature of the alumina by flowing helium gas underneath it. If the procedure is followed properly, a single crystal of sapphire begins to grow from the bottom of the crucible and continues to grow until the whole crucible of melted alumina has solidified into one single crystal. Early sapphire crystals made this way measured two inches in diameter, but crystals now can be made more than 12 inches across. We've come a long way from Verneuil's first success. Single crystals are

Figure 2-7. *Enormous sapphire single crystal. This 13.4 inch diameter, 143 pound sapphire crystal was grown by the heat-exchanger process.* (*Courtesy of Crystal Systems, Inc., Salem, MA.*)

now used not only for gems and bearings but also for lasers, communications equipment, and even the bar code readers in grocery stores.

High-Temperature Furnaces.   Single-crystal growth is pretty exotic and was certainly one of the earliest modern ceramics technologies. The Verneuil work demonstrated that ceramics with special characteristics could be made from pure, chemically synthesized powders, especially if high temperatures could be reached. Achieving high temperature in an enclosed furnace chamber, though, was a much more difficult task than melting particles of powder in a torch. Considerable engineering of furnaces and the development of new ceramic furnace-lining materials (refractories) became a priority during the early 1900s. Another piece of the ceramic puzzle fell into place.

The Automobile.   The spark that really advanced ceramic technology and contributed the next important pieces to the modern ceramics puzzle was the invention of the automobile. Early automobiles, such as those developed by Daimler and Benz in Germany in the 1880s, needed a simple way to ignite the tiny fire that powered their internal combustion engines, much as a match needs to be struck against a special surface to start it burning. Spark plugs provided that magical flash of fire.

A spark plug consists of a ceramic electrical insulator shaped like a tube, with a metal pin inside it and a separate metal case partly surrounding it. The metal case contains a hook that bends around the pin to form what is called an *air gap*. When a car's engine is running a high-voltage spark jumps across the air gap of the spark plug like a miniature bolt of lightning and ignites a mixture of fuel and air inside the engine cylinder. The small explosion that results pushes a piston and provides mechanical power to the engine. These actions create a really severe environment for a ceramic to survive. The ceramic insulator is exposed to high voltage (about 8000 volts), high temperature (bursts to about 4000°F) alternating with cool temperature, and shock waves from the fuel-air explosions. Early 1900s glass and ceramics could not reliably survive these severe conditions. In those early days of the automobile, driving 50 miles was a great adventure, usually resulting in at least one breakdown to replace a spark plug.

Extensive research and development were conducted between about 1903 and 1940 to achieve ceramics that provided longer life and durability in both automotive and aircraft spark plugs. This laid the groundwork for the rapid emergence of modern ceramics. It involved incremental changes in the spark plug insulator composition from a traditional porcelain made with naturally occurring raw materials to high-alumina compositions made with chemically synthesized powders. The transition sounds easy, but it actually required entirely new technologies in powder synthesis, shape fabrication, fabrication equipment, and manufacturing scale-up.

The first porcelain spark plugs were fabricated on a potter's wheel from pottery compositions that had previously been adopted for other types of electrical insulators. The porcelain was fabricated from the naturally occurring raw materials clay, flint, and feldspar. The clay allowed the mixture of powders to become pliable, like potter's clay, when water was added. These spark plugs were not reliable. Researchers determined that additions of alumina powder improved electrical properties and increased durability/reliability. As alumina powder was added and clay deleted, however, the mixture no longer became pliable with the addition of water, so traditional pottery-shaping techniques did not work. Alternate techniques were developed, through which the shape was achieved by adding an organic binder (glue) to the powder and compacting the mixture under pressure inside a shaped cavity, a method similar to the dust-pressing technique developed for tile in England. The powder compact then could be removed from the cavity, machined to a final shape, and fired in a furnace to remove the organic

Figure 2-8. *Automotive spark plug showing some of the metal parts and the white alumina ceramic insulator. (Photograph courtesy of Delphi Automotive Systems, Flint, MI.)*

binder and transform the compact of powders into a dense, strong ceramic. As the years passed, these new processes were refined, and even automated, to meet the high production rate required for automobiles and aircraft. Even by 1920, large quantities of spark plugs were being fabricated. AC Spark Plug, for example, reported producing about 60,000 spark plugs per day just for World War I aircraft requirements. By 1949, over 250 million spark plugs were produced per year worldwide, and by 1997, more than 1 billion per year rolled off the assembly lines worldwide.

Achieving high-temperature furnaces mentioned earlier, was one of the challenges that had to be overcome in spark plug development. Porcelain powder compacts became solid, dense ceramics when heated to around 2382°F (1300°C). However, as the percent of alumina increased, the temperature requirement went beyond the capability of existing production furnaces. For example, 80% alumina required heating to about 2600°F (1427°C), and 95% alumina required a furnace to operate above 2950°F (1621°C). New technology was successfully developed to design and build production furnaces to reach these temperatures, adding another key piece to the puzzle.

amazing facts

The first porcelain spark plugs were fabricated on a potter's wheel!

By 1920, AC Spark Plug was producing 60,000 spark plugs per day for World War I aircraft.

By 1949, over 250 million automotive spark plugs were produced per year. Today, more than 3 million are produced each day worldwide.

Powder Processing.   The final piece of the puzzle was to achieve a method that could produce, at low cost, thousands of tons of pure ceramic powders such as alumina. Large-scale chemical production of alumina powder had been developed by Bayer (the same company that introduced aspirin) clear back in 1888, to supply aluminum oxide raw material for aluminum metal production, but the Bayer powder contained too many sodium impurities to be useful for the new ceramics. Finally, by about 1936, scientists succeeded in producing high-purity, low-cost alumina powder on a large scale.

Alumina Spreads to Other Modern Applications

Once the powder synthesis, composition, and fabrication challenges were solved, high-alumina ceramics became the standard for spark plugs that last for thousands of hours and are one of the most reliable products we can buy. Besides their superior electrical-resistance behavior, high-alumina ceramics also proved to be more resistant to wear, erosion, and corrosion than traditional ceramics. This overall superiority led to a wide variety of products, including abrasives and grinding wheels; wear-resistant tiles for industrial and mining equipment; labware resistant to acids and bases; seals; pump parts; and even bullet-resistant armor. Alumina has been so successful that production reached a level of approximately five million metric tons per year by 1995, enough to make a solid block the size of a football field and nearly 1000 feet high. Pretty incredible!

More than five million metric tons of aluminum oxide products are manufactured in the world each year.

The Explosion of New Ceramics and Ceramic Uses

Alumina was the pioneer of advanced ceramics, the model to be emulated for numerous other advanced ceramic materials that followed. The lessons ceramists had learned from creating products with alumina ceramics were applied to engineering all of the later ceramic compositions. Whole families of ceramics were invented: magnetic ceramics; high-strength, high-temperature ceramics; piezoelectric ceramics; special optical ceramics. "Designer ceramics" soon began to fill every demand of electronics, communications, medicine, transportation, aerospace, power generation, and pollution control.

OVERVIEW: INTO THE FUTURE

What a dramatic impact the evolution of ceramics has had on civilization. In the past 100 years alone, our knowledge of ceramic materials has exploded. We now understand some of the behavior that baffled mankind for centuries. We can design and engineer ceramics, and combine ceramics with other materials, to solve just about any problem or challenge we encounter.

We owe a debt of gratitude to the new advanced, or engineered, ceramics for many of the remarkable products that we now take so much for granted—products such as radios, color televisions, fiberglass, lasers, ultrasonics, microwave and fiberoptic communications devices, advanced jet aircraft, home computers, and cellular phones. We'll discuss these in later chapters, along with important uses of advanced ceramics in the fields of medicine, automobiles, the Space Shuttle, and energy and pollution control. First of all, though, let's look at the aesthetic qualities of ceramics, the magic and beauty that has enhanced human existence for centuries.

The Beauty of Ceramics

CHAPTER 3

*B*eauty has its own kind of magic, and some of the world's most beautiful creations have been crafted in ceramic materials. The exact date and circumstances of the first piece of "ceramic art" are not known, but archaeological discoveries suggest that it was a vessel or figurine hand-formed from damp clay and hardened in the sun. Such items probably were not intended as art but instead as religious articles to assure fertility or successful hunting. In this chapter, we explore the origins and evolution of beauty in ceramics and look at a small sampling of the wide variety of beautiful ceramic art forms.

PRODUCTS AND USES

Decorative tiles
Blown glass
Greek vases
Sculptures, figurines
Chinese porcelain
Islamic glass and tiles
Venetian glassware
Art tiles
Carved glass and cameos
Slip cast ceramic art
Sgraffito
Glass and enamels
Glass flowers
Faceted synthetic gemstones

EARLY CERAMIC ART

Eastern Mediterranean Ceramics

By 5000 BC, decorated pottery was being crafted in the parts of Asia Minor presently occupied by Turkey. Potters decorated this pottery after it had been fired by painting on simple designs with a reddish pigment over a lighter colored clay slip. The pigment and slip were made by mixing the selected color of powdered clay with water to turn the mixtures into thin pastes that could then be smeared or painted onto the surface of the porous pot. Some of the clay particles soaked into the pores with the water, allowing the clay decoration to adhere to the pot. The decorations were geometric patterns, similar to cave wall paintings from the same region.

Decorated pottery also has been found at many sites in ancient Mesopotamia and Egypt estimated to date back to roughly 5000 to 4500 BC. Mesopotamia extended along the valleys of the Tigris and Euphrates Rivers from the east end of the Mediterranean Sea to the Persian Gulf and has sometimes been referred to as the Cradle of Civilization. Egypt was another important early center of civilization along the fertile flood plains of the Nile River. The people from these regions, too, made clay-based earthenware, which they clay painted with simple designs and, occasionally, with animal and human figures. They also used a technique

Figure 3-1. *Musrussu the Dragon. One of the 575 animal figures depicted in glazed tile on the Ishtar Gate, which was constructed in Babylon around 580 BC under the rule of Nebuchadnezzar II.* (Reprinted by permission of the Detroit Institute of Art, Detroit, MI.)

called *incising*, which consisted of scratching a pattern or design into the surface of the pot. For further variation, potters could apply their clay paint into the incised design. If you were to look at some of these old pots in a museum and compare them to the simple, undecorated containers from much earlier years, you would see beauty evolving in the new earthenware.

A big step in the evolution of beauty in pottery coincided with the dawn of the Bronze Age, around 4500 to 4000 BC. As discussed in Chapter 2, bronze could be formed into delicate, thin, contoured shapes. Potters tried to mimic these shapes, opening up new options for beauty in their creations. New kilns that could fire the earthenware at higher temperature were a big help because of the resulting increased strength and decreased porosity. The new two-chamber kilns also opened up more options for decorations. Previous single-chamber kilns had caused any decorations applied before firing to wash out or fade. With the two-chamber kilns, the pots could be isolated in a separate compartment away from the direct flames in the fire box. Clay pigments could now be painted on before firing and would retain their color and sharp edge definition of designs, resulting in increased contrast between the pot and the decoration. This contrast made the decorations more visible and dramatic.

The next spectacular additions to ceramic beauty were glazes and faience, followed by the creation of whole articles crafted from glass. As also discussed in Chapter 2, early glazes emerged from efforts dating back to around 3500 BC to emulate the precious blue stone lapis lazuli. Blue glazes on pottery, tile, and brick were widespread in Egypt, Mesopotamia, and Asia Minor by 2000 BC. Special glazes containing lead were developed sometime between 1750 and 1170 BC in Babylon. The lead acted as a flux, to allow the glaze to form at a lower temperature. Pigments that lost their color at higher temperatures could now be used, resulting in brighter colors and a larger variety of colors. Lead glazes could be applied over the prefired ceramic or even over a higher-temperature glaze, opening up a whole new realm of artistic creativity. By the 8th century BC, another glaze additive was discovered, this time by the Assyrians in Persia. This additive, tin oxide, yielded an opaque white glaze that could actually hide the underlying brown or reddish clay earthenware, so that the ceramic piece appeared much lighter in color. Glazes could now really show off their true colors!

Two famous examples of glazed ceramics in ancient Mesopotamia were the Tower of Babel and the Ishtar Gate built in Babylon under the reign of King Nebuchadnezzar II (604 to562 BC). The Ishtar Gate was about 34.5 feet high and adorned with 575 tin-glazed tile figures of real and mythical animals. The most important animal figure was probably Musrussu the dragon, which served the chief Babylonian god Marduk (Baal). Musrussu was portrayed with the body and head of a serpent, the front legs of a lion, the hind legs and talons of an eagle, and the neck and mane of a bull.

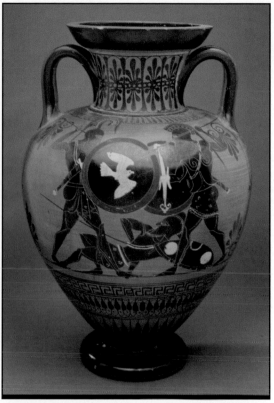

Figure 3-2. Classical Greek vase. This 17 inch high vase is from 510 to 500 BC and was probably used as a storage jar. The side shown depicts a mythical combat scene. The other side depicts Apollo, the god of light and music, standing between his mother and sister, playing a kithara. (Courtesy of the Utah Museum of Fine Arts, Salt Lake City, UT. Photograph by Jim Frankoski.)

Greek and Roman Ceramics

The rise of Greek civilization around 1000 BC had a favorable influence on all of the arts, including ceramics. The Greeks believed strongly that community, religion, and art were closely linked, and art was an integral

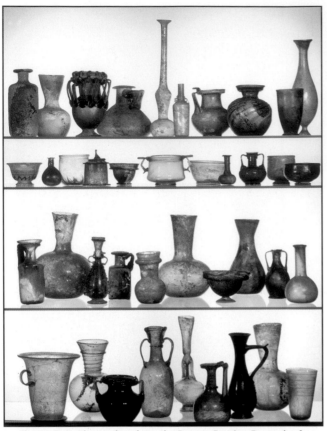

Figure 3-3. *Blown glass from the Roman Empire. Examples from the first through fourth centuries* AD, *showing some of the shapes and colors of glass vessels produced by early glass blowers.*
(Courtesy of The Corning Museum of Glass, Corning, NY.)

part of their daily lives. Greek pottery richly depicted scenes of daily life and mythology. The pots were wheel-thrown and painted with a slip containing very fine particles of clay. Additional detail was achieved by scratching through the painted image to reveal the color of the underlying earthenware. Early pots (about 1000 to 530 BC) had black scenes bordered with decorative geometric and floral designs, all on a red or yellowish background. Later pots (about 530 to 330 BC) presented red figures with other colors on a black background. Each of these Greek styles illustrates a story in a dramatic contrast of colors. Interestingly, the Greek potters chose not to use glazes.

By 250 BC, the classical Greek civilization had begun to decline and was eventually absorbed into Roman civilization. By 150 BC, the Roman Empire completely encircled the Mediterranean Sea and extended north into England. The Romans gleaned ideas and technology from many cultures and spread them widely. One of the most important of these technologies was glassblowing. Early glassblowers became versatile craftsmen. They blew glass vessels in numerous shapes, sizes and colors. Some vessels were blown into the open air and either took a natural (free-form) shape or were contoured with ceramic or metallic tools, while still hot and soft, into other creative forms. Others were blown into shaped molds, so that the glass bubble deformed to duplicate the shape of the inside of the mold. Glass molding opened a wide range of creative opportunities. Some artisans engraved inside the molds elaborate scenes, which were then embossed into the glass during blowing.

Ceramics in the Far East

China was another early center of civilization. Just as in the Mediterranean region, the earliest ceramics were earthenware, and the earliest decorations were painted on with clay or iron oxide (rust) pigments. Produced in the agricultural regions along the Yellow River in northeastern China around 5000 BC, these pots were "coil-built," using washed and strained fine particles of clay deposited by the Yellow River. Potters formed the moistened clay into coils and stacked the coils in rows on top of a flat, circular clay base. As coils were added to the stack, the potter pressed them together by hand and then shaped the pot by tamping or beating the outside with a wooden paddle, while holding a wood block on the inside to retain the container's shape. The surface of the pot was then rubbed (burnished) with a smooth tool, giving the ceramic piece a smooth, polished surface before firing. Variants of this technique are still used today by artisans in many parts of the world.

Early Chinese pots were adorned with spirals, checkered and dotted patterns, fish, frogs, deer, and snakes, but very rarely with human faces. The decorations probably had specific religious or ritual meanings, because many of the early Chinese pots have been found in burial sites, although some decorated urns also have been found at ancient home sites.

Black pottery (from the Longshan Pottery Culture) became popular in China sometime between 3000 and 2000 BC. This pottery seems to have been prepared using a fast-moving wheel and was similar in color, shape, and decoration to earlier black pottery from several Mediterranean regions. Longshan pottery was exquisitely refined and eggshell thin. Chinese ceramists created the rich black color by firing the pots under special "reducing" conditions that involved restricting the amount of oxygen in the kiln by plugging the air holes—much the same way we char a steak by putting the lid on our barbecue grill and closing the air vents. Firing under reducing conditions is still commonly done today to obtain special visual effects.

Figure 3-4. *Neolithic Chinese storage jar. Painted earthenware jar, 11 inches high, from the Yangshao culture, Banshan cultural phase of Gansu Province. Coil-built with buff clay, covered with orange slip, and decorated with black pigment. Decorations limited to the upper portion because the lower portion was designed to be placed firmly into the* **ground.** *(Courtesy of The International Museum of Ceramic Art at Alfred, NY. A gift of Chaoling and Fong Chow, 1992.)*

Just as it had in Europe, the Bronze Age strongly influenced ceramics in China. As mentioned in Chapter 2, ceramic designs began to mimic bronze, and ceramic ware became a low-cost alternative to bronze. High-temperature kilns, special clays such as kaolin, and high-temperature glazes led the Chinese potters along a different path from that followed by Western potters, resulting in an incredible legacy of innovations and ceramic creations unique to the East. The Chinese developed and refined stoneware and porcelain and centuries later exported these ceramics to other countries. On the other hand, they imported glass and lead glaze technologies.

Ceramics in the Americas

The earliest pottery in the Americas was found in Ecuador and is estimated to have originated around 3000 BC. Because North and South America were isolated, ceramics there don't seem to have been influenced

by any other cultures until European explorers arrived after AD 1500. Before that, all of the American pottery was earthenware shaped by hand, decorated with slips, and fired in rather simple kilns. The potter's wheel was never developed in the Americas, and glazes were relatively unknown.

In spite of their limited technology, ancient Americans managed to create highly distinctive decorations, as well as numerous shapes and styles of pottery. For example, the absence of the potter's wheel led them to a greater variety of imaginative, free-form and noncircular shapes than were produced by early Mediterranean and Chinese potters. A popular style was hand-sculpted containers shaped like animals whose heads served as pouring spouts and

Figure 3-6. *Mayan urn. One of two matching urns, dated from* AD *500 to 800, found in a tomb in the central Petén-Tikal/Uaxactun region of Guatemala. This 14.5 inch high earthenware urn depicts a Mayan god with swirling scroll eyes, fish fins on the cheeks, and filed teeth. The frame of beads around the face represents a cave, indicating that the god is in the underworld.* (Courtesy of the Utah Museum of Fine Arts, Salt Lake City, UT. Photograph by Jim Frankoski.)

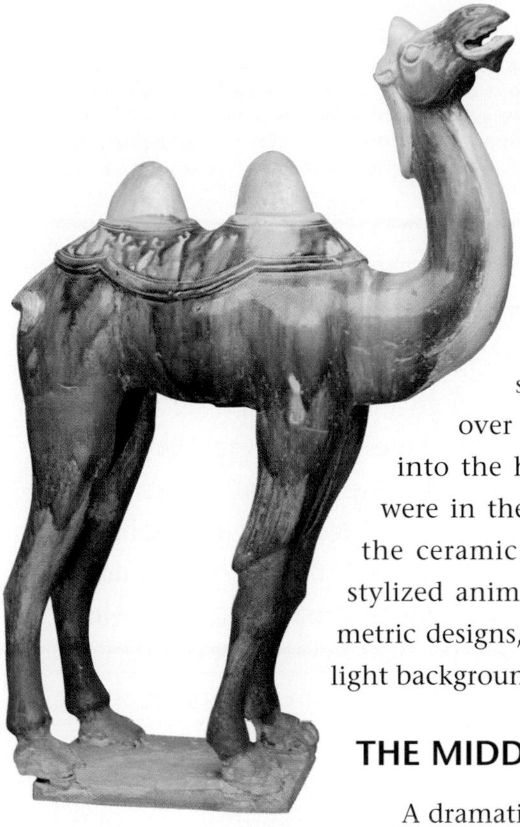

Figure 3-7. *Tang Dynasty camel. Earthenware with chestnut and three-color glazes, this 25.3 inch high camel was buried in the grave of a Chinese aristocrat sometime between* AD *618 and 906. (Courtesy of the Utah Museum of Fine Arts. Photograph by Jim Frankoski. Gift to the museum from Professor and Mrs. Lennox Tierney and the Friends of the Art Museum.)*

legs as supports. Another imaginative style of container had a hollow stirrup-shaped carrying handle arching over the top, with a pouring spout built into the handle. Other earthenware creations were in the images of important gods. Most of the ceramic works were decorated with highly stylized animals, insects, persons, gods, and geometric designs, usually painted in a dark color on a light background.

THE MIDDLE AGES

A dramatic decline in all forms of ceramics and ceramic art in Europe began in about the 5th century AD and continued to about the 11th century. These years made up the period we now refer to as the Dark Ages or Middle Ages. Some stagnation also occurred in other areas of the world between about AD 200 and 600. Even China had an unsettled period between the Han Dynasty (which ended in AD 230) and the Tang Dynasty (which started in AD 618), although stoneware and glazes continued to improve. The Tang Dynasty began many centuries of reasonable stability in China that encouraged the refinement of ceramics. About the same time, Islam arose in the Middle East, leading to a period of stability that also encouraged innovations and creativity in ceramics.

Chinese Ceramics during the Middle Ages

The Tang Dynasty in China, strongly influenced by the spread of Buddhism from India, was a period of peace, tolerance, and substantial interaction and trade with other cultures—a time of great economic prosperity and wealth. It was a perfect environment for ceramic creativity to flourish. Incredibly beautiful ceramic items were crafted, but they were not necessarily considered art at the time by the Chinese. These creations generally had a deeper, spiritual meaning that was an integral part of the Chinese philosophy of life. Today, however, we value Chinese works as art treasures. Some of the most spectacular ceramic items during the Tang Dynasty were commissioned by wealthy aristocrats. Their purpose was to

Figure 3-8. *Chinese imperial vase. Commissioned for use in a Chinese imperial palace around* AD *1736 to 1795, during the Qianlong period. Porcelain, 21.6 inches high, decorated with blue clouds and red bats and coated with a clear glaze. Bats were a common theme. The Chinese words for bat and happiness coincide. The design on the vase was a wish to the viewer, "May your happiness be as high as heaven."* (Vase from the collection of Mr. Bert G. Cliff, courtesy of the Utah Museum of Fine Arts, Salt Lake City, UT. Photograph by Jim Frankoski.)

be buried with the dead during funerals. The camel was a popular subject, based on its importance to the silk trade.

A highlight of the Tang Dynasty was the emergence of porcelain, which later evolved into even greater beauty during the Sung (AD 960 to 1279) through Ming Dynasties (AD 1368 to 1644). Most people have heard of the renowned Ming vases of China. These vases were made of fine white porcelain decorated with intricate scenes and designs that looked almost as if they had been drawn on with ink. The most distinctive ceramic "ink" was made with a special slip containing the chemical cobalt oxide. This slip was delicately painted onto the surface of the porcelain and coated with a glaze. During firing, the cobalt oxide turned the ink a permanent blue, and the glaze became a bright, colorless varnish to accent and protect the scene. Other inks were developed with different colors. Still other beautiful effects were achieved by painting with colored enamels. Enamels were low-temperature glazes that were painted onto the smooth surface of a fired, high-temperature glaze and refired at a low temperature to produce scenes in a variety of colors.

Other innovations in China included the development of slip casting during the Tang Dynasty and of copper-red coloration during the Ming Dynasty. Slip casting involved pouring a fluid slip (a mixture of clay and other ceramic particles suspended in water, similar to the early clay paints) into a porous mold. Water was sucked out of the slip into the pores of the mold, and the ceramic particles in the slip were deposited on the inner walls of the mold. When the desired wall thickness was achieved, the remaining slip was poured out of the mold. After partial drying, the ceramic ware could be removed from the mold, decorated, and fired. The mold was made of bisque-fired ceramic—ceramic fired at a lower temperature than normal, so that it had high porosity. Slip casting did not become a popular fabrication method in China, but centuries later, it was important for the fabrication of European porcelain, and it is widely used today for everything from figurines and beer mugs to sinks and toilets. The second innovation, copper-red color, was achieved by applying a thin wash of copper oxide over a glaze and firing under reducing (low-oxygen) conditions. A whole range of colors, from salmon to pink to blood red, could be produced with that tecnhique.

Figure 3-9. *Glazed tile in the Blue Mosque, built in Istanbul by Sultan Ahmed I around 1603 to 1617. Many of the tiles originally came from the Topkapi Palace, which was partially destroyed by fire in the 16th century. (Courtesy of Kalebodur Seramik Sanayi, A.S. and the Tile Heritage Foundation.)*

Figure 3-10. *Islamic glass mosque lamp. Colorless blown glass, enameled and gilded, about 12 inches high. From Syria, probably Damascus, about AD 1355. The inscription reads, "God is the light of the heavens and the earth. His light is as a niche in which is a lamp, the lamp in a glass, the glass as it were a glittering star". (Courtesy of The Corning Museum of Glass, Corning, NY.)*

Islamic Ceramics during the Middle Ages

Founded between AD 610 and 632, Islamic religion led to a period of stability in the Middle East that paralleled the stability in China. By the middle of the 8th century, a unified Islamic civilization extended from Spain and northern Africa through Egypt, Arabia, and Persia all the way to the borders of India and China, assimilating a rich tapestry of cultures. Because Islamic doctrine prohibited the use of precious metals for tableware, ceramics took the place of gold and silver at the table and also for many religious items. Some of the most beautiful and distinctive glass items and tiles ever made were created by Islamic craftsmen. Decorations on these items were strictly dictated by Islamic law; idolic images of animals and humans were prohibited, so most Islamic ceramic ware was decorated with geometric designs.

Besides their fine glass and tiles, the Moslem countries also made other important contributions to the world of ceramic art. Islamic artisans developed a unique way of working with white tin glazes and cobalt blue pigments to create interesting new decorative effects. They first painted an unfired ceramic piece with the white tin glaze and then applied the cobalt blue pigment over the glaze. When the piece was fired, the blue color would spread into the white glaze, in much the same way that ink spreads out into a piece of blotting paper.

Another Islamic innovation was *lusterware*, produced by painting special chemicals containing gold, silver, or copper onto the surface of either glass or a glazed ceramic. When the ceramic or glass was fired in a strongly reducing atmosphere (high fuel, very low oxygen), the chemical decomposed to release the metal as a thin, lustrous coating on the surface of the glass or ceramic, similar to the shiny metal coating we apply to the back of mirrors.

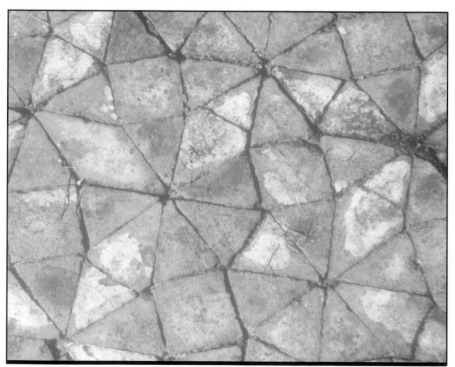

Figure 3-11. *Ceramic floor tiles from 12th century England. From the high altar at Fountains Abbey, North Yorkshire, England. Begun in 1132 by 13 young monks of the Cistercian Order.* (Courtesy of the Tile Heritage Foundation. Photo by Joseph Taylor.)

Figure 3-12. *Mid-14th century floor tiles. Penn Tiles from Buckinghamshire, England, laid on the floor of the Pitstone Church. Among the earliest tiles commercially produced in England and commonly used on the floors of parish churches, manor houses, and merchants' houses.* (Courtesy of the Tile Heritage Foundation. Photo by Joseph Taylor.)

Yet another interesting technique refined by Islamic craftsmen was *sgraffito*, a variation of incising. Sgraffito is created by applying a clay slip over a different color clay and selectively scratching away the surface layer to reveal the contrasting clay underneath.

THE RENAISSANCE AND BEYOND

The Dark Ages finally came to a close in Europe. Interest in ceramic art rebounded, especially within the Church. By the 12th century, decorated tiles were commonly used on the floors of parish churches, manor houses, and merchants' houses. Byzantine Gothic churches were resplendent with brightly colored mosaic tiles and glazed floor tiles. Brilliant stained-glass windows in Gothic and Renaissance churches recreated Biblical scenes with mosaics of colored glass, through which the "light of Christ" filtered down

Figure 3-13. Stained glass window. Stained-glass window from a church in Oxford, England. (Photo by D. Richerson.)

onto the heads of devout worshipers. As economic prosperity returned to Europe, patronage of the arts flourished, ushering in beautiful ceramic creations to go along with the unsurpassed artistic works of Michelangelo, Leonardo da Vinci, and Raphael.

By AD 1300, a major glass industry had arisen in Italy, centered on the island of Murano off the coast of Venice. Venice was the major sea power of the day and the leading trade center between East and West. Glass became an important item of trade. Venetian glassmakers absorbed and adapted the skills of the Islamic glassmakers and produced an impressive variety of stunning glass creations, ranging in colors from colorless (*cristallo*) to emerald green, dark blue, amethyst, reddish brown, and milky white. They introduced marbleized glass, refined techniques for enameling and engraving, and even made high-quality mirrors. Venetian craftsmen dominated the world market for fine-quality glassware for several centuries.

Figure 3-14.(top) *Maiolica floor tiles. A magnificent display of maiolica floor tiles in the Cappella di San Sebastiano, Basilica of San Petronio, Bologna, Italy; tiles produced in 1487 by Pietro di Andrea in Faenze, Italy.* (Courtesy of the Tile Heritage Foundation. Photo by Nicola Di Nunzio.)

Figure 3-15. *Late 17th century Portuguese tile panel. Measures 57 by 78.75 inches and shows carpetlike pattern popular with Islamic craftsmen and adopted by Spanish and Portuguese craftsmen.* (Photo from the Tile Heritage Foundation, Courtesy of the Snug Harbor Cultural Center, Staten Island, NY.)

The quality of their work was so superb, and their secrets thought to be so valuable, that glassmakers who tried to leave Murano were subject to the death penalty! Fortunately for the rest of the world, some Murano glass artisans did escape, and the Venetian glassmaking secrets slowly spread across Europe and England.

INFLUENCE FROM THE MOSLEM WORLD

The influence of Islamic ceramics migrated northward to enhance the resurgence of ceramic art in Europe. By 1400, Spanish potters had refined the tin glaze and lusterware techniques from the Moslem world, creating spectacular dishes, vases, jars, and other decorated ceramic items. The tin-glaze technique was adopted later in Italy, where it became popularly known as *maiolica*. A magnificent display of maiolica floor tiles is in the Basilica of San Petronio in Bologna, Italy. Maiolica tiles often were combined with frescoed walls and ceilings and with sculpted or hand-painted beams.

Influence from the Far East

The tales of Marco Polo and other visitors to China stimulated interest in porcelain and attempts to duplicate it as early as 1300. Many new forms of ceramic ware were created that mimicked some aspects of the appearance of porcelain: bone china, translucent mixtures of glass and ceramic particles, whiteware with blue scenes.

Figure 3-16. Wedgwood Porcelain. Modern samples, showing the classical style that was introduced in the 18th century in England. (Photo by D. Richerson.)

None of them, however, duplicated porcelain until the work of Ehrenfried Walther von Tschirnhausen and Johann Friedrich Böttger at Meissen in Saxony, Germany. The Meissen artisans didn't just try to duplicate Chinese porcelain objects, however. Instead, they created their own designs and styles. They focused on European and Christian themes and on making three-dimensional porcelain sculptures depicting people, animals, flowers, and objects. The Meissen craftsmen made great strides in the arts of slip casting and mold making and in joining together multiple pieces before firing them into a single complex, ornate work of art.

Josiah Wedgwood (1730 to 1795), in England, took a different approach with porcelain. Rather than producing delicate, three-dimensional artistic creations, he focused on making porcelain dinnerware and other household items that could be sold to everyone, not just the wealthy art collector. His cameo style, which consisted of a raised scene on a different-colored background, was a big hit and is still manufactured today. Wedgwood even introduced an assembly line approach to the manufacturing of porcelain.

Figure 3-17. *"Comet" pattern pressed glass. Produced in New England and Pittsburgh areas around 1850. Tallest piece measures 13.7 inches.* (Courtesy of The Corning Museum of Glass, Corning, NY. Entire grouping a gift to the museum in memory of Amy Chace.)

Pottery and Glass in Colonial North America

European and English pottery techniques were introduced to North America during the early years of colonization. Glassblowers were among the Jamestown settlers in 1607, although the first successful glass factory didn't appear until 1739. Earthenware was produced starting in about 1640, stoneware in about 1720, and porcelain in 1770. Styles and techniques were traditional English and European. When the United States was formed and a federal government established by 1789, a tariff was imposed on imported products, including ceramics. This tariff stimulated the start-up of additional potteries, but pottery styles continued to be European, with no major new artistic styles or techniques introduced.

Glass was a different story. To meet the demand for quantity with some degree of beauty, early U.S. glassmakers mimicked the appearance of cut glass by blowing glass into a shaped mold. By the 1820s, though, they came up with a revolutionary new technique to get the same cut-glass appearance. Rather than blowing the glass into a mold, the glass makers mechanically pressed the glass into the mold. This exciting, new innovation, was perhaps one of the most important since the invention of glassblowing during the Roman Empire. Two relatively unskilled workers could produce four times as much glassware as three or four skilled glassblowers had been able to make before. The new pressed glass was as attractive as mold-blown glass, and because it was easier to make, its beauty became more affordable to the average person.

Figure 3-18. *Mid-19th century encaustic floor tiles. From Saint Giles' Roman Catholic Church, Cheadle, Staffordshire, England. Designed by noted architect Augustus Pugin and produced by the famous English tile manufacturing pioneer Herbert Minton. Decorative floor tiles became popular throughout the British Empire and in America.* (Courtesy of the Tile Heritage Foundation. Photo by Joseph Taylor.)

The Influence of the Industrial Revolution

The Industrial Revolution had both positive and negative effects on the aesthetic qualities of ceramics. From a positive perspective, the Industrial Revolution introduced new mechanized equipment that could produce decorative ceramic items in far greater numbers than ever before, making them available to a larger cross section of society. Dust pressing of tiles and pressing of glass

Figure 3-19. *Dust-pressed Victorian art tile. Produced by the United States Encaustic Tile Works, Indianapolis, Indiana, around 1890. Victorian tile depicting romanticized scenes, wildlife, and other realistic subjects were produced in large quantity at factories in England and the eastern and midwestern United States and were important decorations in homes and buildings of the late 19th century.* (Sample from a private collection. Courtesy of the Tile Heritage Foundation. Photo by Joseph Taylor.)

are examples of the new industrial methods. Decorated and undecorated tiles became so inexpensive during the second half of the 1800s that they were used in almost every imaginable way in buildings, homes, hospitals, and even dairies.

When the machines took over, however, craftsmen and their hand-made ware could no longer compete. Ceramic art began to stagnate. Then in about 1850, a major movement began in Europe to preserve the art of crafting ceramic and glass articles for their aesthetic value. This Arts and Crafts Movement lasted into the 20th century and resulted in the founding

Figure 3-20. *Cut-glass plate. "Russian" pattern, crafted in 1906 by T. G. Hawkes & Co., Corning, NY. Measures 13.4 inches in diameter. A large service of this same pattern was prepared for the White House in 1891.* (Courtesy of The Corning Museum of Glass, Corning, NY. Plate was a gift of T.G. Hawkes & Co.)

Figure 3-21. *(Left) The Great Dish. Exquisite cameo dish carved from five-layered glass by George Woodall and team at Thomas Webb & Sons, Amblecote, England. The base layer was dark green, followed by white, yellow-green, white and red.* (Courtesy of The Corning Museum of Glass, Corning, NY. From a bequest of Mrs. Leonard S. Rakow.)

of numerous art schools, small pottery studios, and museums. The democratic philosophy of the movement was that art should be made by the people, for the people, and for the enjoyment of both the maker and the user. Ceramic crafting became very popular again, leading to almost limitless individual experimentation with techniques and styles. That attitude has continued intermittently to the present time.

A CENTURY OF CERAMIC ART

The past century has indeed been a time of unlimited experimentation by individuals and even within large companies. All of the techniques refined through the centuries are now accessible to virtually anyone willing to take the time and effort to study them. Artists and crafters can use their imaginations to style their own unique ceramic creations. The remainder of this chapter offers a small sampling of some of the interesting styles and creations of the past century or so.

Classical to Abstract Glass

Glass art at the turn of the century was strongly influenced by the Victorian Age. Large, extravagant creations in cut glass were highly prized. The finest quality of colorless lead-crystal glass, which has incredible transparency and brilliance, was carefully cut and polished to create plates and other objects that appeared to be covered with faceted diamonds.

Another classical style of glass was cameo, which dates back to at least the time of the Roman Empire. Exceptional cameo glass plates and other objects were carved during the Victorian Age, near the end of the 1800s. The glass was prepared in layers, each with a different color. Carving into the glass layers, the skilled artist could create a picture that resembled a three-dimensional painting and even had subtle shadings between the colors of the layers. Classical mythological (especially Greek) and religious themes were popular.

A completely different style of glasswork was crafted by Leopold Blaschka and his son Rudolf, of Hosterwitz, Germany. The Botanical Museum of Harvard University com-

Figure 3-22. Glass cactus and flower. Scientific replica of Echinocereus engelmanii, crafted completely from glass by Leopold and Rudolph Blaschka in 1895. (Courtesy of The Botanical Museum of Harvard University, Cambridge, MA. Photograph by Hillel Burger.)

Figure 3-23. Glass tableware. Transparent cobalt blue glass produced in about 1916 by blowing. Designed by Josef Hoffmann, Weiner Werkstatte, Vienna, Austria. Tallest piece measures 12.9 inches. (Courtesy of the Corning Museum of Glass, Corning, NY.)

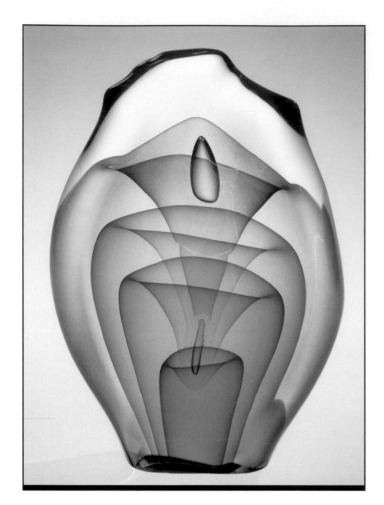

Figure 3-24. *Glass sculpture, "Emergence Four-Stage."* Contemporary sculpture from 1975, by Dominick Labino, using colorless and pink transparent glass with amber tints. Measures 8.8 inches high. *(Courtesy of The Corning Museum of Glass, Corning, NY. Purchased with aid of funds from the National Endowment for the Arts.)*

missioned them to prepare detailed, botanically accurate, three-dimensional glass replicas of plants for use in botany classes. Between 1887 and 1936, the Blaschkas crafted nearly 3000 life-sized glass models illustrating over 830 species of plants, including carnivorous plants, as well as insect pollination, and plant diseases (fungi) and life cycles. They even crafted magnified cross sections and enlarged plant parts. This remarkable collection has been preserved as the Ware Collection of Blaschka Glass Models of Plants at the Botanical Museum of Harvard University. Interestingly, the Blaschkas traced a continuous family line of glassmakers back to a 15th-century Venetian glassworker.

Traditional glass tableware has continued to fluorish throughout the world during the past century. Pay a visit to your local Mikasa store or to boutiques and antique shops, and you'll see an astonishing array of simple and highly decorative glass. You may not even have to go that far: You probably have a pretty good assortment in your own cupboards and china cabinet.

Contemporary glass artists have explored highly imaginative and abstract expressions in glass. Because glass can be worked in a flowing molten state, it can be swirled and contoured into infinite patterns and designs. Colors, voids, and streaks can be worked in for further variety. The glass can be cooled and resoftened, pulled, pressed, contoured with a tool, pressed or poured into a shaped mold, partially or

Figure 3-25. *Press-molded glazed Chinese Shiwan Gongzai stoneware.* **Modern piece from Foshan, China. Measures 22 inches high.** *(Photograph courtesy of the Director of Shiwan Artistic Ceramics Factory, Foshan, People's Republic of China, provided by Ed McEndarfer, Truman State University, Kirksville, MO.)*

completely crystallized into a polycrystalline ceramic, and bonded while hot to other pieces of glass. It can be cut and polished, and three-dimensional images that appear as beautiful sculptures when viewed through the glass from the other side can be carved carefully from one side into its interior. Beautiful gradations of color, such as pink shading into purple or blue to green, are possible. The possibilities for creating beauty in glass seem almost unlimited.

Expressions in Pottery

Pottery is crafted throughout the world yet today. In some regions, such as southern China, pottery for daily use and trade is still made by the same techniques passed down from generation to generation for centuries. Whole towns are dedicated to the production of pottery using the wheel and other techniques, such as press molding. For example, in Foshan, People's

Figure 3-26. *Charles Fergus Binns glazed stoneware vase. Wheel-thrown vase, 1908.* (Courtesy of the International Museum of Ceramic Art at Alfred, Alfred, NY.)

Figure 3-27. *Contemporary stoneware teapot and vase. The neck of the vase and the teapot were shaped on a potter's wheel. The rest of the vase, which is 13 inches high, was slab-built. Made with Sacramento graystone clay with a white tin oxide glaze that allowed aesthetic streaks of iron stain to come through.* (Courtesy of the artist, Donaree Neville, Salt Lake City, UT. Photograph by Brad Nelson.)

Figure 3-28. *Contemporary Santa Clara Pueblo pottery. Two pieces of coil-built, carved earthenware by Tammy Garcia. The blackware jar is 16 inches high, was fired outdoors in traditional fashion, and won the Best of Pottery award at the 1995 Indian Market in Santa Fe, New Mexico. The red pot, "Pueblo Girls," is 19 inches high, was fired in an electric kiln, and won first place in the nontraditional category.* (Courtesy of Tammy Garcia, Taos, NM.)

Republic of China, press-molded stoneware is made in a tradition that dates back to the Ming Dynasty. Beautiful figurines are formed by pressing clay into shaped plaster molds, joining pieces of clay together, decorating with slips and glazes, and firing.

Figure 3-29. *Slip-cast and glazed earthenware. Contemporary work by Dan and Nisha Ferguson illustrating spectacular use of slip casting and glazing. The piece measures 15 by 15 by 15 inches.* (Courtesy of Dan and Nisha Ferguson, Toronto, Ontario. Photograph by Vivian Gast.)

Stoneware also has been popular among studio potters in the United States. A pioneer in the early 1900s was Charles Fergus Binns. He was the founder and director of the New York State School of Clay Working and Ceramics from 1900 to 1931 and is fondly given the title Father of American Studio Ceramics. Binns also guided the emergence of advanced ceramics and the transition of the School of Clay Working and Ceramics to later became the New York State College of Ceramics at Alfred University, one of the leading schools of ceramic science and engineering in the world.

The Binns stoneware vase in Figure 3-26 was crafted on a potter's wheel. Contemporary studio potters use the wheel as well as other techniques such as *slab*

Figure 3-30. *Porcelain platter with sgraffito decoration. Contemporary plate, 18 inches in diameter, crafted by artist Wayne Bates.* (Courtesy of Wayne L. Bates, Murray, KY.)

building. A versatile technique, slab building uses a flat sheet of clay, similar to dough that has been rolled out to make sweet rolls or a pie shell. The clay sheet is cut to the desired shape and stuck to other pieces of clay at the edges to construct the stoneware article. For further variation, slabs can be linked to wheel-thrown sections.

Figure 3-31. *Silicon carbide gravy boat. Fabricated by Shenango Refractories, The Pfaltzgraff Company, by slip casting. Measures 7.8 inches long. Originally displayed at a manufacturers' exposition to attract attention and show the shape-forming capabilities of the manufacturer. Donated in 1997 to The International Museum of Art at Alfred and featured in the exhibition Conspicuous Applications of Advanced Ceramics, curated by Dr. William Walker.* (Courtesy of the International Museum of Art at Alfred, Alfred, NY.)

Artists continue to create with earthenware. Some of the potters in the southwestern United States dig their own clay and use a combination of ancient and modern techniques to shape it. Tammy Garcia, a fifth-generation potter in the Santa Clara Pueblo Native American tradition, creates pots using the coil-built method, with clays dug near the Santa Clara Pueblo. Her designs are produced by carving, followed by burnishing to achieve a smooth, bright sheen. She sometimes fires her pots outdoors in traditional fashion but also uses an electric kiln, taking advantage of the special effects that can be achieved with both old and new techniques.

Some of the most spectacular contemporary earthenware creations are formed by slip casting. Dan and Nisha Ferguson of Toronto, Canada use slip casting and bright glaze colors to craft earthenware with a circus theme. Dan sculpts models of the animals, which he then uses to make multipiece plaster molds. He slip casts the animals and bowls. Nisha designs the patterns on the bowls and renders them with glazes. The bright colors are achieved through primary- and secondary-colored underglazes, sometimes requiring as many as five separate firings.

Today's potters also create beautiful ceramics with porcelain. Wayne Bates links the ancient art of sgraffito with porcelain to craft gorgeous, large platters. He forms a platter on an electric wheel, coats the platter with brightly colored engobes (similar to a slip, but with less clay), and carves by the sgraffito technique to selectively reveal the underlying white porcelain. Next, he bisque-fires the piece and then applies a glaze with a satin texture before a final high-temperature firing.

Figure 3-33. *Synthetic gemstones mounted in beautiful jewelry. Fine example of Inamori jewelry, named in honor of Dr. Inamori, who founded the Kyocera Corporation and was a key leader in the development of advanced (fine) ceramics in Japan.* (Courtesy of Kyocera.)

Figure 3-32. *Synthetic gemstones. Synthetic ruby weighing 120 carats and surrounded by five synthetic star rubies and sapphires, two synthetic emeralds, and three diamond imitations.* (Courtesy of Kurt Nassau, author of the book Gems Made by Man, Chilton Book Co., Radnor, PA.)

Art from Modern Ceramics

As scientists develop new ceramic materials, artists wait at the doors of the science laboratories to latch onto these new ceramics and create new expressions of art. One artist has carved the honeycomb ceramic invented for the pollution-control systems of cars into driftwoodlike shapes and other forms, especially for use in decorative lamps. A company that makes ceramics for high-temperature furnace linings and other ultra-high-temperature applications has slip cast a gravy boat from silicon carbide. Silicon carbide is not a naturally occurring material on earth; it has only been found in some meteorites. Ceramic engineers have learned to synthesize silicon carbide, so that we now even have art indirectly from outer space.

A final example of the use of advanced ceramics to make beautiful creations is synthetic gemstones. Scientists now know how to synthesize rubies, sapphires, star rubies and sapphires, emeralds, amethysts, garnets, diamonds, and even opals. Some new manmade gemstones, such as cubic zirconia, are as brilliant as diamonds—but at a fraction of the cost. These synthetic gemstones are now crafted into jewelry that looks fit for a king or queen but is affordable to just about everyone.

OVERVIEW: ENDURING BEAUTY

Ceramics and their magical beauty have truly enhanced people's lives since the earliest civilizations. Hopefully, the beauty of ceramics will never fade from our world. Books that you might want to read for more information about the timeless beauty of ceramics are listed at the end of this book. If you ever get to Corning, New York, be sure to visit the Corning Museum of Glass. If you're in the Boston area, don't miss the Botanical Museum of Harvard University Ware Collection of glass flowers. Truly incredible!

Ceramics and Light

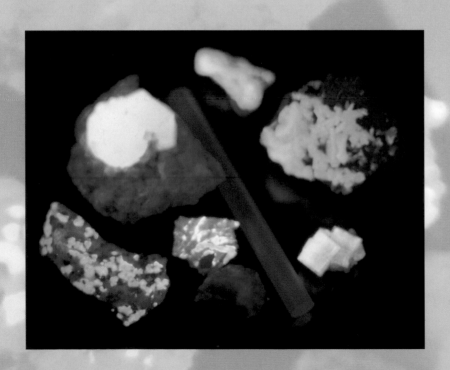

CHAPTER 4

The previous chapter showed us many spectacular examples of how beautiful ceramics can be. But where do they get their beauty? Why do they have such bright colors? Why are many ceramics transparent or translucent, while metals are so opaque that we can't see even a trace of light through them? The answers to these questions are all based on the magical interaction between ceramics and light, called optical behavior. Let's look first at the two most basic optical properties that we see every day—transparency and color. Then we'll examine some more exotic optical behaviors, such as phosphorescence and electroluminescence, that we also see every day but are less familiar with.

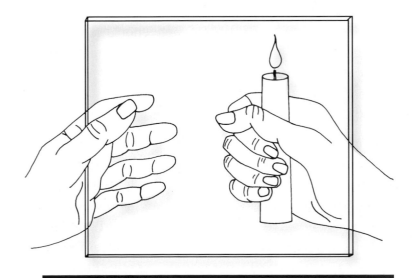

Figure 4-1. Transparency, the magic of seeing through a solid object. *(Illustration by Angie Egan, Salt Lake City, UT.)*

Figure 4-2. Red Canyon near Bryce Canyon, UT, viewed on a cold winter day through the transparent glass of a car window. (Photograph by D. Richerson).

TRANSPARENCY, A RARE AND SPECIAL QUALITY

We usually think of a transparent object as one that we can see through clearly, a substance through which light can pass without any difficulty. Not many natural solid materials are completely transparent. Natural crystals like quartz are often transparent, but these crystals typically are too small to be used to make actual transparent products. About the only exception is mica, those pearly flakes in granite that can be peeled off in thin, flexible layers. Some mica crystals are large enough and transparent enough that thin sheets of them were once used as small windows before flat sheets of glass became available.

Light and Electrons

The combination of conditions that allows a material to be transparent and that affects most of its other optical properties involves the interaction of light with the electrons in the material. To understand this, we need to briefly review what an atom looks like and how atoms bond with each other to form a solid material. We also need to understand a little about light. First, let's look at atoms. Each atom has tiny electrons circling a nucleus of protons and neutrons like the planets circling the sun. Some electrons are in orbits near the nucleus, like Mercury and Venus. Others are farther from the nucleus, like Jupiter and Neptune. The electrons farthest from the nucleus are especially important in determining the behavior of a material. They dictate how atoms can bond together and whether the resulting combination of atoms will have the characteristics of wood, water, metal, plastic,...or ceramic. Let's compare the way the atoms bond together—and the role of the electrons—for metals, plastics, and ceramics and see how this behavior affects the interaction of these materials with light, especially in terms of why metals are opaque and ceramics and plastics can be transparent.

First, let's imagine what the inner atomic structure of a metal or ceramic looks like. For simplicity, think of each atom as a miniature solar system with lots of open space between the electrons and the nucleus but with an overall spherical shape like a ball. These atoms are arranged side by side and stacked layer upon layer, much the same as if we carefully filled a box with rubber balls. For a metal, some of the electrons in the outer orbits of each atom are free to move into similar outer orbits in the adjacent atoms and on into other atoms throughout the whole piece of metal. This free movement of these electrons throughout the metal is the glue that bonds the metal together as well as the source of many of the properties we associate with a metal, such as conduction of electricity and heat—and the opaque appearance, as you will see shortly.

Now that we can visualize the stacking of the atoms, the open space, and the movement of the electrons in a metal, we can take the next step to understanding the optical behavior of a metal or any other material: reviewing the nature of light and how it interacts with a material. Light is a form of energy. It can be pictured in a couple of different ways. One picture depicts light as tiny waves, similar to the ripples on a pond but much smaller; the other describes light as tiny packets of energy called *photons*. These light waves or photons are so tiny that they can interact with the electrons in a material. Under the right circumstances, electrons can actually absorb (soak up) the energy of the light, so that the light can't pass through the material. To do this, however, an electron must be able to add this extra energy to its own energy. The only way the electron can do this is to move out of its usual orbit and into a higher-energy orbit farther from the nucleus. That's where the characteristics of the material come in. The bond electrons are free to move anywhere throughout a

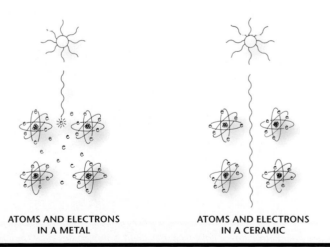

ATOMS AND ELECTRONS
IN A METAL

ATOMS AND ELECTRONS
IN A CERAMIC

Figure 4-3. *Light interacts with electrons in a material to determine whether the material is transparent or opaque. Some electrons in metals are able to add energy from a light beam to their own energy and move to a higher-energy orbit. The light is absorbed and cannot pass through, so the metal is opaque. The electrons in most ceramic or glass materials are not free to interact with the light and move to a higher-energy orbit, so cannot absorb the light energy; the light is able to pass through, and the material is transparent.* (Illustration by Angie Egan, Salt Lake City, UT.)

metal and can easily pick up extra energy and move to a higher energy orbit. Thus, they can readily absorb the light energy. Since the light energy is absorbed, it does not pass through the metal, and the metal appears opaque and reflective.

Why are ceramics and plastics different? The atoms are bonded together differently in these materials than in a metal. Rather than being free to roam throughout the material, the electrons that bond together glass or plastic or a single crystal of a ceramic are tied up tightly between two neighboring atoms and have no new orbit available to easily move into. They can't absorb the light energy, so the light passes through the material, which then appears transparent. Polycrystalline ceramics would also be transparent except that the features of the microstructure (grains, grain boundaries, pores, and inclusions) that we discussed in Chapter 2 usually block some of the light and scatter the rest in many directions. That's why even most ceramics appear opaque or translucent rather than perfectly transparent.

Refraction—Where's the Fish?

An important characteristic of transparent materials is *refraction*. Even though the electrons in a transparent material do not absorb the light energy, they still interact slightly with the light, to slow it down as it passes through the material. Slowing down causes the light to change direction, just as we observe when light enters water. We've all reached into water for an object or fish and been surprised that it was not where we visualized it would be. This is because the water slowed down the light and bent, or *refracted*, the light beam. You can easily demonstrate refraction in your kitchen. Dip a pencil or straw at an angle into a glass of water, and look from the side so you can see both above and below the water. The pencil or straw seems to change direction slightly, right where it enters the water.

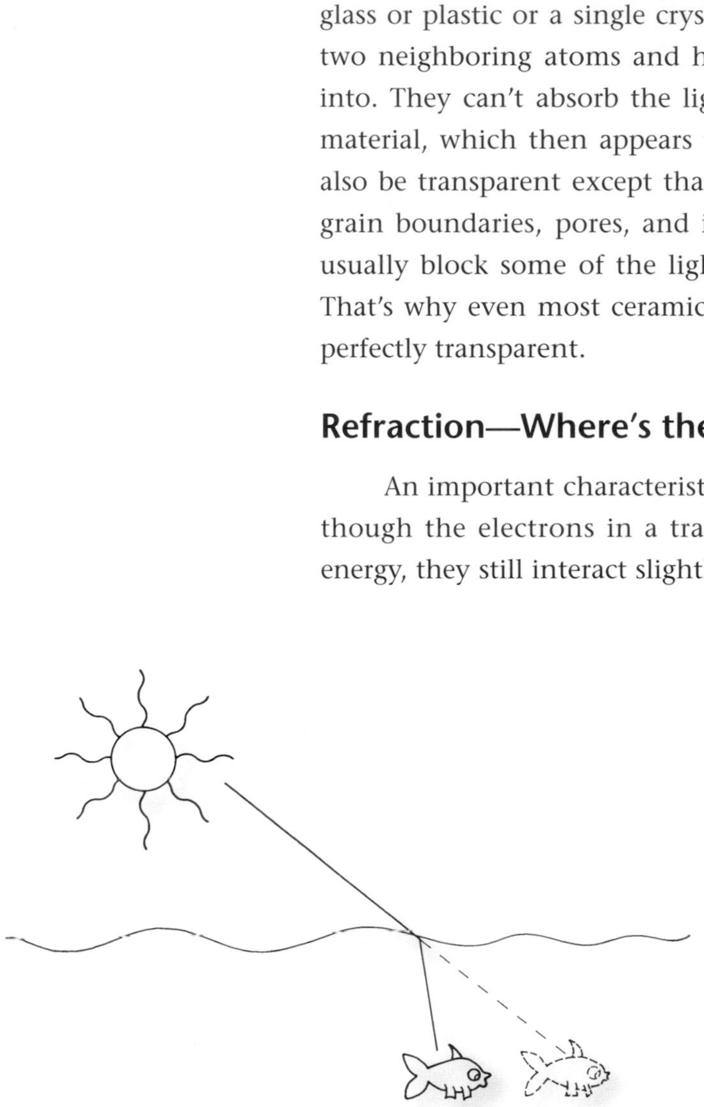

Figure 4-4. *Refraction of light by water. Light is slowed down by the water so that the fish appears to be in a different place than it actually is.* (Illustration by Angie Egan, Salt Lake City, UT.)

All transparent ceramics and glass refract light, just as water does, but some refract it more than others because the different ceramics interact with light differently. Diamond, for example, refracts light much more than does window glass. In fact, a faceted diamond refracts the light so much that the light bounces around inside the gemstone and emanates from it with beautiful brilliance and fire, which is one of the reasons that diamond is so highly valued as a gemstone. For years, scientists have searched for a transparent ceramic that refracts like diamond but costs less. Their closest imitation is a synthetic gemstone called *cubic zirconia*, which we'll discuss later in this chapter, when we talk about color.

Transparency in Action

Our most important and widely used transparent ceramic is glass. Because it can be fabricated as a molten fluid, glass can be formed into an enormous variety of shapes and sizes for thousands of products; yet most of us take glass for granted. How often do you look through a window or your eyeglasses and think how really special and magical their transparency is? How often do you look at a light bulb or your television screen or your computer monitor and realize that none of them would be possible without the special characteristics of glass? Glass is so important to us that about 600 million tons are produced worldwide each year, mostly for products that require transparency. I could devote a whole book just to glass.

Many practical uses for glass, such as windows, fiberglass insulation, and containers were identified in Chapter 1; and many beautiful applications

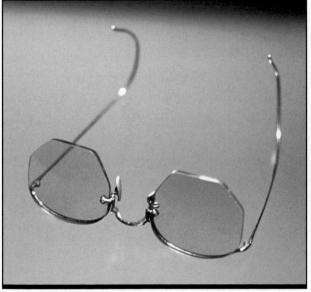

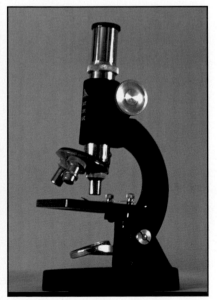

Figure 4-5. (Left) *Eyeglasses help us to see the world more clearly.* (Photograph by D. Richerson.)

Figure 4-6. (Right) *The glass lenses in a microscope help us to see the life in a drop of water.* (Photograph by D. Richerson.)

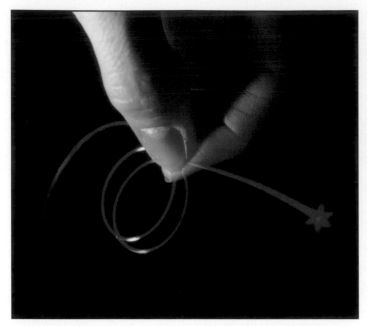

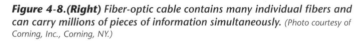

Figure 4-7.(Left) *Light can travel around corners in optical fibers.* (Photo courtesy of Corning, Inc., Corning, NY.)

Figure 4-8.(Right) *Fiber-optic cable contains many individual fibers and can carry millions of pieces of information simultaneously.* (Photo courtesy of Corning, Inc., Corning, NY.)

were reviewed in Chapter 3. Other important and fascinating uses for glass in telescopes, automobiles, and even medicine are revealed in later chapters. But one of the most important ways that our modern world relies on the transparency of glass is for fiber-optic communications.

Our world has become dependent on rapid communications, whether from our house to our neighbor's or from our country to another. Until about 1976, these communications were accomplished by sending electrical messages through copper wire. Only a limited number of messages can be sent simultaneously through copper wire, however, so myriad strands of copper wire were needed to send large numbers of messages. Thousands of miles of copper-wire-filled cables, about as big around as the CDs for your stereo disk player and weighing millions of tons, were snaked across the oceans' floors and across the continents in an effort to meet the ever-increasing demand.

During the 1950s and 1960s, communications scientists discovered that tremendous quantities of information could be encoded in light and transmitted through glass fibers thinner than a human hair. The glass fibers and light sources available at that time, however, could only transmit the information about 10 feet (less than a flashlight can project in air). Even though the glass in these early fibers was transparent, tiny bubbles and irregularities scattered the light and prevented it from traveling through any longer lengths of fiber. Two major inventions made fiber-optic communication feasible. One breakthrough was the invention of the laser by T. H. Maiman, in about 1960, providing a light source that could shoot out a beam of light intense enough to travel thousands of miles. The second innovation was the development by 1970 of optical-quality glass fibers by Drs. Robert Maurer, Donald Keck, and Peter Schultz at Corning, Inc. These fibers could carry the light from a laser for three-fifths of a mile, but they still lost about 99 percent of the light intensity because of imperfections in the glass. However, that was good enough to get fiber-optic technology off the ground. Since 1970, glass-fiber technology has been refined so much that fibers now can retain 1 percent of a light impulse for more than 75 miles.

Transparency and the Electromagnetic Spectrum

We've discussed how the electrons in ceramics interact with light to affect optical properties, but light is only one of many energy sources that interacts with ceramics.

Light—visible light—is a form of energy called *electromagnetic radiation.* Electromagnetic radiation is produced when another electron in an atom jumps from one orbit to another orbit closer to the nucleus. This radiation travels in what are called *electromagnetic waves,* so-named because the waves have both an electric and a magnetic component. Electromagnetic waves are capable of transporting energy in such a way that they don't need anything like air or water to move from one place to another—they can even travel through a vacuum, which is completely empty space.

To imagine electromagnetic waves, picture waves just like those you see at the ocean or the lakeshore. Each light wave has a high point, or *peak,* followed by a valley, or *trough,* before the next wave starts. No matter what type of wave is involved, the distance from the peak of one wave to the peak of the next wave is called the *wavelength.* For waves in the ocean, this distance is large and can be measured in feet or meters. Most electromagnetic waves, however, are incredibly tiny, and their wavelengths are measured in very small units of length called nanometers. "Nano" is the Greek word for *one-billionth;* one nanometer equals one-billionth of a meter (a meter equals slightly more than three feet). It takes over 25,000 nanometers just to equal the thickness of a human hair.

amazing facts

- Two optical fibers, each thinner than a hair, can transmit 625,000 telephone calls at once.

- About one-quarter *pound* of optical fiber can transmit as much information as nearly two and one-half *tons* of copper.

- Optical fibers are so strong that a 1 inch diameter cable could lift 216 elephants, each weighing 6 tons.

- Since 1976, enough optical fiber has been installed in the world to stretch back and forth between the Earth and the moon 160 times.

(Source: Corning, Inc., Corning, NY.)

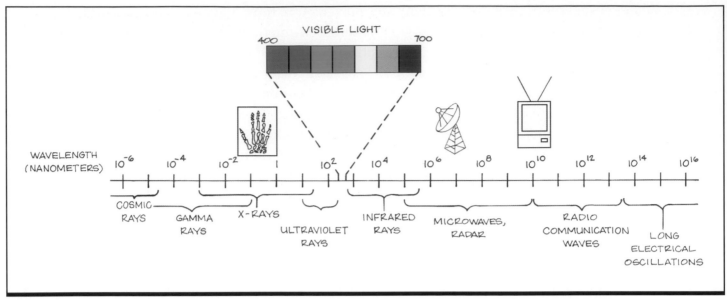

Figure 4-9. *The electromagnetic spectrum. Light that is visible to our eyes is a very small portion of the electromagnetic spectrum. (Illustration by Angie Egan, Salt Lake City, UT.)*

The further an electron jumps between orbits, the more energy is present in the electromagnetic radiation and the shorter its wavelength. When you line up all the different electromagnetic waves according to their respective wavelengths, you have a wide band that scientists call the *electromagnetic spectrum.* Waves on one end of the spectrum—like radio waves—have longer wavelengths (and less energy), while waves on the other end—like gamma rays—have shorter wavelengths (and more energy). For example, gamma rays, which are created by nuclear explosions, have tiny wavelengths, less than 10 trillionths of a meter, and 10 billion times the energy of visible light. Exposure to gamma rays can kill you. Ultraviolet rays have larger wavelengths (and less energy); their damaging rays can cause sunburn. Infrared rays, which carry information from your remote control to your television and are important in night-vision devices, have even larger wavelengths. Finally, at the other end of the electromagnetic spectrum, radio waves, which are used to transmit radio and television signals, have wavelengths several meters long.

Light that can be seen by our eyes is called *visible light.* Visible light is located roughly in the middle of the spectrum, from about 400 to 700 nanometers (a distance about one fiftieth of the thickness of a hair), and is an incredibly small portion of the overall electromagnetic spectrum, as illustrated in Figure 4-9. We'll discuss visible light in more detail soon.

Even though we can't see them, the electromagnetic waves outside the visible range are important to us. We depend on radio waves to transmit radio and television signals, as well as signals for pagers and cell phones.

The same microwaves that heat our food in microwave ovens are used for TV and communications satellites and radar. X-rays (which have fairly small wavelengths) have enough energy to penetrate our skin. Gamma rays are used in medicine to kill diseased cells and in the scientific study of materials.

These other electromagnetic waves travel through materials similarly to the way light does. For a material to be transparent to a certain wave, it must have bond electrons that can't absorb the wave's energy. A good example of this need arises in communications with an airplane, missile, or spacecraft. For the same reasons that metals are opaque to visible light, they also are opaque to ultraviolet and larger wavelength rays, including microwaves and radio waves, which are used to send communications to these craft. If a whole craft is built of metal, the communication wavelengths can't get through. The vehicle requires an *electromagnetic window* that's transparent to whatever wavelength is being used for communications. Because some glass and polycrystalline ceramics are transparent to microwaves and radio waves, they are built into the structure of an aircraft, spacecraft, or missile to act as an electromagnetic window. For a missile, the window is called a *radome* and is typically a cone-shaped covering on the tip of the missile that houses the guidance system. The radome doesn't have to be transparent to visible

Figure 4-10. Ceramic radomes. *(Photo Courtesy of Ceradyne, Inc., Costa Mesa, CA).*

light, only to microwaves or radio waves. But the materials in missiles and airplanes do have to be resistant to some pretty severe structural challenges like rapid changes in temperature, and impact with rain drops at speeds well over 1000 miles per hour. Running into a rain drop at 1000 to 3000 miles per hour can do a lot of damage to a ceramic, so ceramic engineers have developed special ceramics to meet these needs. The Space Shuttle also has special requirements for windows. In addition to being transparent to communications wavelengths, resisting very high temperature during re-entry, and surviving rain-drop impact, the Space Shuttle windows must also be transparent to visible light, so that the pilot can see through the window to land the shuttle.

COLOR

Color is certainly one of the most magical and beautiful attributes of any material. Like transparency, color is another optical property caused by the interaction of light with the electrons in a material.

Where Does Color Come From?

Take a look again at Figure 4-9 on page 66. A beam of white light (the kind that's all around us, that we receive from the sun) actually contains all of the colors of the rainbow, and each color has its own unique wavelength. As long as all of the wavelengths are mixed together, the light appears white. If this white light falls on a piece of glass or ceramic and all the wavelengths pass through, the glass or ceramic appears transparent and colorless. If the ceramic is polycrystalline and all of the wavelengths are scattered equally, the ceramic will appear white and either translucent or opaque. If the white light falls on a metal, all of the wavelengths will be absorbed or reflected, and the metal will appear opaque. Color arises when some of the visible wavelengths are absorbed and the rest are either allowed to pass through or are scattered.

Let's look at some examples and then at a few different ways that ceramics are able to filter out some wavelengths and let others through.

Wavelength (nanometers)	Spectral Color of Wavelength	Complementary Color
410	Violet	Lemon-yellow
430	Indigo	Yellow
480	Blue	Orange
500	Blue-green	Red
530	Green	Purple
560	Lemon-yellow	Violet
580	Yellow	Indigo
610	Orange	Blue
680	Red	Blue-green

Figure 4-11. The colors of light waves.

When a single wavelength or narrow band of wavelengths is absorbed by a material, our eyes will see the color of the remaining wavelengths that pass through it or are reflected back. For example, if the orange color of the spectrum (wavelength of about 610 nm) is absorbed, the material will appear blue. If green is absorbed, the material will appear purple. The color we see is referred to as the *complementary color*. The spectral colors, their wavelengths, and their complementary colors are shown in Figure 4-11.

Color in Ceramics

How are ceramics able to absorb some wavelengths and allow others to pass? The answer involves the way the electrons in the ceramic interact with the light. As we discussed before, pure ceramics have all of their electrons tied up between nearest-neighbor atoms and have no higher energy orbits for electrons to move into, so they can't absorb extra energy from light. However, by adding small amounts of certain other atoms as impurities, we can modify the ceramic so that some higher-energy electron orbits will be open if just the right amount of energy is added. This "just right" amount of energy corresponds to the energy of a narrow band of wavelength of light.

Cubic zirconia is produced at 5036°F.

Cubic zirconia gems were introduced in 1976.

13 tons of cubic zirconia were produced in 1980 and faceted into about 13.5 million carats of fine gemstones.

Over 500 tons of cubic zirconia were produced in 1997.

Let's look at an example. Remember in Chapter 2 how we mentioned that the red color in ruby comes from a small addition of the chemical element chromium to aluminum oxide? Well, chromium modifies the alumina just enough so that the blue-green wavelength (about 500 nm) from white light can cause some electrons to jump (or *transition*) to a higher energy orbit, thus absorbing the blue-green light waves. All of the other wavelengths pass through, so the ruby appears to our eyes as the complementary color, red. Other impurities make electrons transition when exposed to different wavelengths. Cobalt additions filter out orange wavelengths in some materials to give the "cobalt blue" that was so important in Islamic glazes and Chinese porcelain decorations. Iron additions filter out indigo and green to give colors ranging from yellowish to rust.

Ceramists have learned through the years the right combinations of additives to come up with just about any color we can imagine. Sometimes the ceramic is crushed into a powder and added to other materials as a colorant or pigment, especially when the material or object is created or used at a high enough temperature that other types of pigments are destroyed. Porcelain enamels, glazes, glass, and synthetic gemstones all require ceramic colorants.

Color in Cubic Zirconia. A good example of the use of ceramic colorants is cubic zirconia, which has become a popular synthetic gemstone.

The cubic zirconia crystals from which the gemstones are cut, or *faceted*, are grown from molten zirconium oxide at about 5036°F, where any other type of colorant would be destroyed. A wide range of colors has been achieved in cubic zirconia by adding elements with electrons that absorb different combinations of wavelengths of white light. Vanadium results in green, cerium in yellow to orange to red, iron in yellow, cobalt in lilac, and europium in pink. The many colors of cubic zirconia imitate the precious

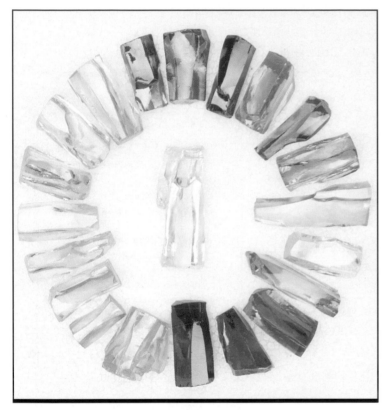

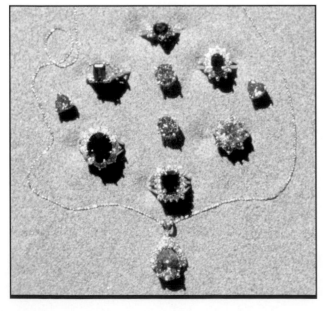

Figure 4-12.(Above) *Rough single crystals and faceted gemstones of cubic zirconia. Crystals grown by Ceres Corporation, Waltham, MA.* (Photographs reprinted with the permission of Dr. Kurt Nassau, author of the book Gems Made by Man, Chilton Book Co., Radnor, PA.)

Figure 4-13.(Left) *Jewelry crafted with cubic zirconia gemstones.* (Photograph by D. Richerson.)

and semiprecious gemstones we love so much: brilliant, colorless diamond, loaded with fire; purple amethyst; red garnet and ruby; and even green emerald. Because of the high refraction of cubic zirconia, these imitations actually outsparkle the natural gems.

Color in Semiconductor Ceramics. Color in ceramics isn't created by impurities alone. Another way color occurs is through a family of

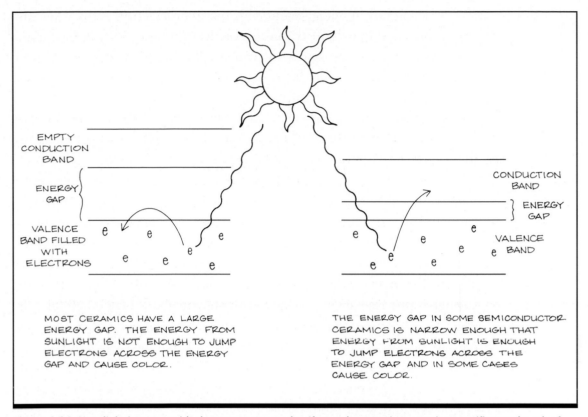

EMPTY
CONDUCTION
BAND

ENERGY
GAP

VALENCE
BAND FILLED
WITH
ELECTRONS

CONDUCTION
BAND

ENERGY
GAP

VALENCE
BAND

MOST CERAMICS HAVE A LARGE
ENERGY GAP. THE ENERGY FROM
SUNLIGHT IS NOT ENOUGH TO JUMP
ELECTRONS ACROSS THE ENERGY
GAP AND CAUSE COLOR.

THE ENERGY GAP IN SOME SEMICONDUCTOR
CERAMICS IS NARROW ENOUGH THAT
ENERGY FROM SUNLIGHT IS ENOUGH
TO JUMP ELECTRONS ACROSS THE
ENERGY GAP AND IN SOME CASES
CAUSE COLOR.

Figure 4-14. *How light interacts with electrons to cause color. If enough energy is present in a specific wavelength of light to cause electrons to jump across the energy gap from the valence band to the conduction band, that wavelength is absorbed. The remaining wavelengths in the light pass through and determine the color of the material.* (Illustration by Angie Egan, Salt Lake City, UT.)

materials called *semiconductors*. To understand what a semiconductor is and how it affects color, we need to review the actions of electrons in a little more depth, especially how electrons jump, or move, between energy levels (which are called *bands*) in a material. Remember those electrons that were involved in bonding the atoms of a material together? We call those bond electrons *valence electrons* and their orbits within the atom the *valence band*. When atoms bond together, a new band forms called the *conduction band*. In a metal, this conduction band overlaps with the valence band, which is why the bond electrons can move (be conducted) throughout the metal. In most ceramic materials, the conduction band is a long way from the valence band—that is, there is a large energy gap between the the two bands. This is called the *band gap*. Adding impurities, as we discussed before, builds a sort of pipeline between the valence band and the conduction band. Semiconductors, though, don't need a pipeline. Semiconductors have a smaller band gap than most ceramics, which means that less energy is required to jump an electron from the valence band to the conduction band. Sometimes, the energy in a certain wavelength of light is just enough, as illustrated in Figure 4-14.

The amount of energy required to jump an electron across the band gap is measured by scientists in units of electron volts (eV). Pure alumina has a band gap greater than 8 eV, which is too large for the energy from light to jump an electron all the way across the gap into the conduction band. The band gap of semiconductors, however, is much smaller. For example, cadmium sulfide has a band gap of only 2.45 eV. The blue and violet wavelengths (from 410 to 500 nm) in light have enough energy to jump electrons across the 2.45 eV gap into the conduction band, thus absorbing these wavelengths of the light. The larger wavelengths in the light (from about 500 to 680 nm) don't have enough energy and thus pass through the cadmium sulfide, so we see the complementary color yellow. In contrast, silicon (Si) has a band gap of only 1.1 eV and thus requires less energy for electron transitions. All of the visible wavelengths of light are absorbed, and the silicon appears opaque and metallic. As an interesting sidelight, this interaction of light with silicon is the magic behind solar cells. The electrons stimulated by the light to the conduction band in silicon can be harnessed to provide electricity, just like the solar cells that produce the electricity to run your solar calculator. Such *photovoltaic* electricity is abundant whenever the sun is shining, so it doesn't waste any natural resources and doesn't produce any pollution.

PHOSPHORESCENCE

Some Ceramics Glow in the Dark

The next time you visit a natural history museum, ask to see the fluorescent mineral exhibit. The mineral samples are typically in a small,

Figure 4-15. *Naturally occurring ceramic materials and a ruby laser rod that all exhibit phosphorescence. The rock and mineral samples and the laser rod on the left are shown in normal daylight; the same samples are shown on the right exposed to ultraviolet light.* (Photos by D. Richerson.)

darkened room or an enclosed case with a window through which you can peek. Usually, you can push a button to see what the specimens look like under normal daylight. Then you push another button, the lights go out, and an ultraviolet light (also called a *black light*) comes on. The rock and mineral samples immediately glow bright green, red, white, and other colors. This magical transformation is an example of *fluorescence,* which is a form of *phosphorescence.*

Phosphorescence is the general term we use when a material gives off light when it is stimulated, or excited, by an appropriate energy source. In the case of the fluorescent rocks and minerals, the energy source was ultraviolet wavelengths of light. The energy from the ultraviolet light in such a case is absorbed by specific electrons in the material. These electrons jump to a higher energy band, but then immediately drop back to their original energy state. As they drop back, they give up the energy they initially absorbed, but in the form of photons, those tiny packets of light energy we discussed earlier. This light has a specific wavelength, which corresponds to a specific color. The material glows in the dark with the emitted light and is referred to as a *phosphor.*

Phosphorescent Ceramics at Work

Ceramic phosphors are very much a part of our everyday life. We're exposed to the light emitted by ceramic phosphors whenever we walk into a room with fluorescent lights or watch television. The fluorescent light consists of a sealed glass tube coated on the inside with a ceramic phosphor. The interior of the sealed glass tube is filled with a mixture of mercury vapor and argon gas. When we turn on the electricity, an electric discharge inside the tube causes the mercury vapor to emit light, but this light is at a smaller wavelength than we can see. Even though we can't see it, this wavelength is easily absorbed by the phosphor, which then emits a broad band of visible white light. The

Figure 4-16. Fluorescent lights, which are lined with a thin inner coating of ceramic that produces light by phosphorescence. (Photo by D. Richerson.)

fluorescent light has several advantages over a conventional incandescent light. The incandescent light produces light by passing electricity through a thin filament of tungsten metal, causing the tungsten to heat up to such a high temperature that it gives off light, just as a hot fire or the sun gives off light. Incandescent bulbs are hot, too bright and glaring to look at, and use a lot of electricity. Fluorescent lights give off soft, cool light and use much less electricity. They also last much longer.

Figure 4-17. *The color in your television comes from the glow of ceramic phosphors.* (Photo by D. Richerson.)

Television also depends on ceramic phosphors. The inside of the screen of the television set is coated with three different phosphorescent ceramics that are stimulated by an electron beam, which rapidly sweeps back and forth across the screen as a signal comes into the set from your cable or satellite dish. One of the phosphors glows red, one green, and one blue. All of the colors of the rainbow can be formed by different combinations of these three colors. The phosphors are carefully designed so that the glowing color fades, or decays, about one-tenth to one-hundredth of a second after the electron beam passes. This allows the next image to be formed quickly, so that the TV has high resolution and the illusion of live action.

The radar screens necessary to monitor the positions of aircraft, and the sonar screens used to search for underwater objects, also use ceramic phosphors. These phosphors are designed for the glow to fade slowly. Most of us have seen this effect in nautical movies. Remember how a bright line sweeps across the ship's sonar screen from left to right, leaving a slowly fading image behind it, and then shows up on the left side to sweep across the screen again?

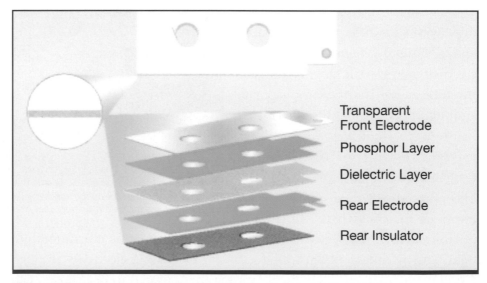

Figure 4-18. *The thin layers that are stacked to form an electroluminescent lamp.* (Courtesy of the Durel Corporation, Chandler, AZ.)

Electroluminescent Lighting

Many ceramic phosphors emit light when they're sandwiched between two sheets of other materials that conduct electricity—that is, when they're placed in an electric field. The light produced in this way is commonly referred to as *electroluminescent lighting,* or simply EL. Widely used for lighting the keypads of cellular phones, the instrument panels of automobiles, and the faces of watches (such as the Indiglo Watch from Timex), EL lamps are thin, flexible, and don't give off heat.

Figure 4-19. *Electroluminescent lighting for TIMEX Indiglo watches. (Photo courtesy of TIMEX.)*

EL lamps are about as thick as the cardboard in a cereal box and consist of a sandwich of different layers of material stuck together to form a laminate. The base layer is made of transparent polyester plastic. A very thin layer of a ceramic called *indium tin oxide* is deposited onto the surface of the plastic. Indium tin oxide is unique because it is, at the same time, both electrically conductive and transparent, a very rare combination. A layer of a ceramic phosphor next is deposited onto the indium tin oxide, followed by another ceramic that's a good electrical insulator (prevents electrons from passing through). The fifth layer is another material that conducts electricity, usually carbon. The final layer is more plastic, another electrical insulator.

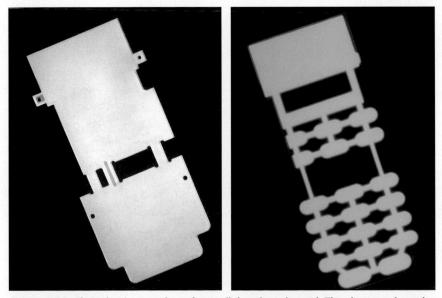

Figure 4-20. *Electroluminescent lamp for a cellular-phone keypad. The photograph on the left shows the multilayer EL lamp that is installed into a cellular phone to illuminate the keypad; the photograph on the right shows the EL lamp with electricity applied. (Photos courtesy of Durel Corporation, Chandler, AZ.)*

During use, an alternating electric current applied to the two electrically conducting layers (electrodes) of the EL sandwich produces an electric field across the phosphor, which stimulates the phosphor to emit light. The electrical power needed to achieve the light is very low: For example, a wristwatch light operates off a 1.5 volt battery at a current of only 3 mA (three-thousandths of an ampere), whereas a cellular-phone keypad requires only a 6 volt battery and 40 mA and consumes about one-quarter of a watt of electrical power. For comparison, a single outdoor Christmas light bulb consumes about 7 watts of electrical power, and a light bulb for a bedside lamp typically uses 40 to 60 watts. Have you ever seen the television commercial in which the lights go out in the middle of a football game? One by one, the spectators turn on their cellular phones, lighting the whole stadium with an eerie green light. This is a novel example of the use of electroluminescent ceramics.

Light from Semiconductors

Another important source of light driven by an electrical input is the *light-emitting diode,* commonly referred to as an LED. LEDs are made of a thin layer of a semiconductor material, such as gallium arsenide, gallium phosphide, or gallium nitride, on a ceramic substrate, such as synthetic sapphire. Red and yellow LEDs have been available for years. After a long-term quest, blue and green LEDs were finally achieved within the last few years. A typical LED is tiny, measuring only a few millimeters across (about one-half the thickness of a pencil), but emits a surprisingly bright light with a very small electrical input (20 mA).

amazing facts

The worldwide market for LEDs is currently 30 to 40 billion per year.

LEDs are used for the on/off lights in many electrical devices such as cellular phones. Lately, they are used more and more in traffic signals, where an array of fewer than 50 LEDs provides the same amount of light as that put out by a light bulb, but with less energy. As a further advantage, LEDs last around 50,000 hours, which saves an enormous amount of maintenance. Outdoor billboards are another rapidly growing use for LEDs, especially in Japan, where billboards have been constructed with millions of individual LEDs. These billboards look like enormous TV screens, but they have better resolution.

LASERS

LASER is actually an acronym for *Light Amplification by the Stimulated Emission of Radiation.* The first laser was developed by T. H. Maiman and reported in 1960. The key component of this laser was a rod of single crystal ruby, which happens to be a ceramic phosphor. As you learned in Chapter 2, ruby is single-crystal aluminum oxide containing a small amount of chromium oxide. The alumina is referred to as the *laser host* and the chromium as the *dopant.* When exposed to a high-intensity flash of light from a flash lamp, electrons in the chromium dopant absorb some of the energy from the light and momentarily jump to a higher energy state. They then drop back into their initial energy state, emitting a single wavelength of light unique to the chromium dopant. The wavelength for the chromium-doped alumina is 694 nanometers, which is dark red. This alone does not make a laser, however.

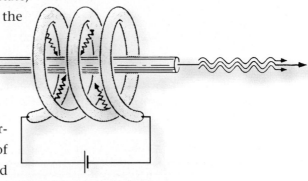

Figure 4-21. *Schematic of the original ruby laser.*

Maiman took laser technology an important step further: He polished the two ends of the ruby rod, so that they were very flat and smooth, and coated one end with silver, which made that end act as a mirror to any light produced inside the rod. Maiman also positioned a mirror at the other end of the rod. With each flash of the lamp, light was emitted within the ruby rod, but it couldn't escape. Bouncing back and forth between the mirrors at the ends of the rod, the light built up intensity. When the mirror on one end was moved, a high-intensity pulse of red light escaped. This amazing pulse of light was different from any pulse of light that had ever been seen before. A normal light beam, such as we get when we shine a flashlight or our automobile headlights, spreads out and completely fades away in just a short distance. The ruby laser light didn't spread out but instead magically remained a single piercing beam for an incredible distance. Why was this ruby laser light so different? The answer—the secret behind the magic—was that the ruby laser light was *monochromatic* and *coherent,* as illustrated in Figure 4-22. The light emitted by the ruby laser was all one wavelength, and all of the waves were lined up one on top of another, so that they reinforced each other to produce a very intense beam.

The laser is one of the most important inventions of the 20th century. Since 1960, lasers have been used in thousands of ways: for fiber-optic communications, cutting, heat treating, guidance systems, measuring devices, holography, bar code readers, entertainment (like laser shows), CD readers, and medicine. Each of these many applications for lasers requires specific laser characteristics that have been made possible by the discovery, since 1960, of a wide range of ceramic hosts and dopants. As an example of a special laser for a particular use, one type of eye surgery involves removing blood on the inside of the eye and stopping the bleeding. The laser used must emit a light beam in a wavelength that is only absorbed by blood, but does not injure surrounding tissue. This feat is achieved with an argon gas laser that emits a specific wavelength of green light. Although the argon laser doesn't have a ceramic laser host, it does have a ceramic core tube through which the argon flows. Besides guiding the flow of the argon, the ceramic performs as an electrical insulator, and also removes excess heat.

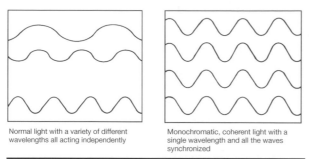

Normal light with a variety of different wavelengths all acting independently

Monochromatic, coherent light with a single wavelength and all the waves synchronized

Figure 4-22. *An explanation of the terms* monochromatic *and* coherent.

Laser engineers have dreamed for years about a ceramic laser that could emit more than one wavelength. Recently, they achieved their goal using a sapphire host doped with titanium. This laser, which can be tuned to emit a broad range of infrared wavelengths (from 700 to 1100 nm), represents another major breakthrough in the field of optical ceramics and is expected to be especially valuable to the medical profession.

MORE OPTICAL MAGIC

After the magical interactions with light that we've already discussed for ceramic materials, it's hard to believe there are additional amazing interactions. Some ceramic materials, though, are *electro-optic,* with refraction

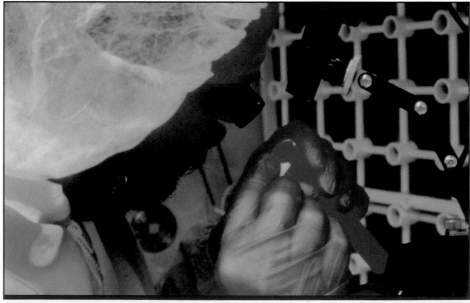

Figure 4-23. *Ruby laser rod being inspected.* (Photograph courtesy of Union Carbide Crystal Products, Washougal, WA.)

Figure 4-24. Single crystals and laser rods cut from them. *(Photo courtesy of Union Carbide Crystal Products, Washougal, WA.)*

characteristics that change when an electric field is applied; others are *acousto-optic* and change refraction behavior when a pressure or acoustic (sound) wave is applied.

Electro-optics

Electro-optic ceramics are beginning to find use in fiber-optic communication systems. Sending information long distances by fiber-optics requires many of the same functions as sending information electrically: amplification, switching, and modulation, among others. Initial fiber-optic systems couldn't accomplish these functions optically, so they had to convert each time to an electrical signal, which was very inefficient. Electro-optic technology now has been developed to perform amplification, switching, and other functions optically. In fact, this exciting new field of *photonics*

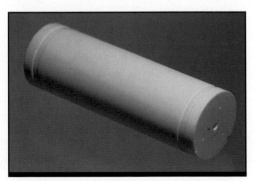

Figure 4-25. Ceramic core tube for a gas laser. *(Sample provided by General Ceramics, Haskell, NJ. Photo by D. Richerson.)*

offers promise for major breakthroughs in fiber-optic communications and in many other technologies previously dominated by electronics.

Figure 4-26. *Pilot wearing electro-optic goggles in the off position.* (Photo courtesy of Sandia National Laboratories, Albuquerque, NM.)

Figure 4-27. *Electro-optic lens in the on position, showing the high level of transparency.* (Photo courtesy of Sandia National Laboratories, Albuquerque, NM.)

An early use for electro-optics, demonstrated in the 1970s, was as military flash-blindness goggles for Strategic Air Command B-52 and FB-111 pilots. Combat pilots are continually at risk of temporary blindness or permanent damage to their eyes from a brilliant flash of light such as a nuclear explosion. The key ceramics technology that made flash-blindness goggles possible was the development of a transparent, polycrystalline electro-optic ceramic known as PLZT (which stands for *lead lanthanum zirconate-titanate*) in 1969 by G.H. Haertling at Sandia National Laboratories. The PLZT for those goggles was engineered into a lens that allows light to pass through when a specific electric field is present. If a bright flash is detected by photodiode sensors (sensors that detect light and send out an electrical signal), the lens shuts off in less than 150 microseconds (1000 times faster than you can blink), preventing damage to the pilot's eyesight. However, the pilot still needs vision to fly the aircraft, so the PLZT lens is designed to instantaneously readjust after the flash to let through only the amount of light that was there before. The pilot thus retains continuous, complete vision. This magical behavior sounds more like science fiction than fact!

Photosensitive Ceramic Materials

Some glass compositions are *photosensitive* and change chemical characteristics when light hits them. The most widely used photosensitive glass is *photochromic glass,* which darkens when exposed to sunlight and fades back to colorless indoors. This capability may not seem remarkable to us today, but it involves some very unique chemistry that the alchemists only dreamed about many years ago: creating a precious metal—in this case silver—"out of nowhere." Let's explore how this mysterious act of magic is accomplished.

Glass is like a universal solvent; almost any group of atoms can be dissolved at high temperature in glass, similarly to the way that sugar or salt can be dissolved in water at room temperature. Photochromic glass contains atoms of silver and halogens (such as bromine and fluorine). The silver and halogens dissolve in the glass during melting, but they combine during cooling into tiny crystals of silver halide ceramic about 5 to 30 nm across that are uniformly distributed throughout the glass. The crystals are much smaller than the wavelength of visible light (400 to 700 nm) and so do not block or scatter the light. As a result, the glass is transparent. However, when bright sunlight shines on the glass, each tiny silver halide droplet absorbs energy from some of the ultraviolet and violet wavelengths of the light to fuel a tiny chemical reaction. This reaction deposits a tiny speck of pure silver metal. The reaction and the presence of metallic silver absorb a portion of the light, so the glass darkens. When the bright light source is removed, as when the person wearing the glasses goes back indoors, the silver recombines with the halogen to form transparent silver halide, and the glass fades back to colorless. Photochromic glass was developed by Armistead and Stookey at Corning Glass in the early 1960s and has been popular ever since as eyeglass lenses.

Other combinations of atoms can be dissolved in glass to obtain a variety of

Figure 4-28. *Images permanently imprinted in photosensitive* **glass.** *(Photo courtesy of the Corning Incorporated Department of Archives and Records Management, Corning, NY.)*

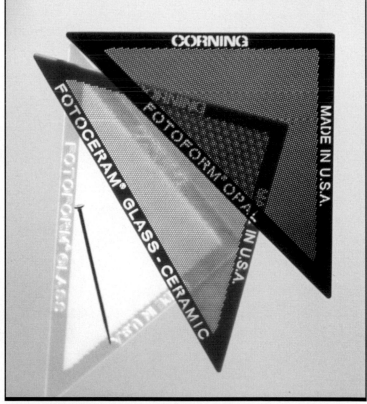

Figure 4-29. *Glass and glass ceramic parts containing thousands of tiny holes produced by the Fotoform process. Straight pin included for size comparison. (Photo by D. Richerson.)*

interactions with light. One combination makes the glass sensitive to exposure to ultraviolet light, so that it darkens later when heated. The amount of darkening depends on how long or how intensely the glass was exposed to the ultraviolet light. This idea has been used to create an image such as a picture directly in glass. A photographic negative is placed over a flat plate of the glass, and ultraviolet light is shined through the negative onto the glass. Darker areas in the negative restrict the amount of ultraviolet light that reaches the glass, while lighter areas let most of the light through to the glass. When the glass is then heated to the appropriate temperature, the picture magically appears, permanently frozen in the glass. Another combination of atoms added to the glass can produce a whole spectrum of colors by selected ultraviolet light exposure and heating. These photosensitive ceramic combinations are called *polychromatic glasses*. Engineers haven't yet found a lot of uses for these novel ceramics, but they're still looking.

Figure 4-30. *Close-up view showing the uniformity and tiny size of the holes produced by the Fotoform process. Straight pin included for size comparison.* (Photo by D. Richerson.)

Figure 4-31. *Additional shapes made using the Fotoform process.* (Photo courtesy of the Corning Incorporated Department of Archives and Records Management, Corning, NY.)

Another photosensitive glass, though, has led to a variety of uses. This glass can be covered with a mask (an opaque material, such as paper or metal foil, with a pattern of holes or designs) and exposed to ultraviolet light. Only the glass under the holes in the mask "sees" the light. When the glass is then heated to the appropriate temperature, its chemistry changes where it was exposed to the light. This region becomes about ten times more soluble in acid than the unexposed glass and can be readily dissolved away, leaving a piece of glass with holes duplicating the pattern of holes in the mask. Much more complex patterns of holes have been created than could possibly be formed by drilling or other techniques. One example is a screen with 300 uniform holes per linear inch, which is about the same as punching 80,000 holes in a postage stamp. Developed by Don Stookey and co-workers at Corning, this photosensitive, etchable glass is referred to as *Fotoform*.

Corning has developed several variants to the Fotoform process: One results in a clear glass; another can be heat-treated after etching to convert the glass to a polycrystalline ceramic named *Fotoceram;* a third ends up as a mixture of glass and tiny dispersed crystals and is identified as *Fotoform Opal*.

OTHER OPTICAL APPLICATIONS

We've covered a lot of territory in this chapter and discussed many fascinating ways that ceramic materials interact with light to make our world a better place; but we easily could have selected many different examples to discuss. Most of the examples we've covered involve only optics for the very narrow band of visible light. Many other optics are also important for the infrared, ultraviolet, and other unseen wavelengths. For example, ceramic materials are critical for night vision devices and other forms of infrared (heat) sensing devices used for mapping temperature distributions, such as in the fields of medicine, astronomy, and Earth studies by satellites.

We've completely ignored major breakthroughs in manufacturing transparent ceramics, especially glass. An important example is the *float-glass process,* a method for making thin, smooth sheets of glass. The molten glass is flowed over an ultrasmooth pool of molten tin to form a continuous glass sheet more than 10 feet across. A single plant can produce 600 tons of sheet glass per day. This process has dramatically decreased the cost of glass and has led to a tremendous increase in the use of glass in modernistic architecture.

OVERVIEW: VISION INTO THE FUTURE

What can we look forward to in the future? Fiber-optics combined with electro-optics will continue to be a part of our lives and certainly will dominate the field of telecommunications. Optical fibers and lasers will be used routinely for many more new medical procedures that replace major surgery with simpler, safer procedures. Perhaps doctors will even be able to clean cholesterol deposits from the insides of our blood vessels using optical technology. Some of these optical technologies are already progressing rapidly, but others are still in the imaginations of scientists and engineers or at early stages of development. One of the most exciting of these new visions of the future is optical computing, which we will use light rather than electrical circuits for computers beyond our wildest imagination.

Amazing Strength and Stability

W e often think of ceramic materials as weak and easily broken. One of our earliest memories may be of getting into trouble for breaking a glass, a dish, or a window. Experience has taught us that handling articles of glass or ceramics requires extra care. But is this perception of weakness really true? What about brick and concrete? They're ceramic, too, and are used to construct buildings, bridges and highways. What about glass fibers for fiber-optics and for reinforcing plastics? Remember in the previous chapter, when we mentioned that a 1 inch diameter cable of glass fibers could theoretically support the weight of 216 six-ton elephants? That makes at least some ceramic materials stronger than steel!

PRODUCTS AND USES

Silicon nitride jet-engine parts

Transformation-toughened zirconia that can be beaten with a hammer without breaking

Glass and carbon fibers for the ultra-lightweight construction of modern airplanes and space vehicles

Silicon nitride ball bearings that last three to ten times longer than metal bearings

Silicon nitride tools that can cut metals five times more rapidly than any prior tool material can cut

Ceramic engineers have made remarkable breakthroughs in just the past 30 years to come up with new ceramics that are stronger than we ever imagined possible. They also have invented new ceramic materials that are incredibly stable, can survive in severe environments that destroy other materials, or can perform tasks that other materials can't. Let's start this chapter by reviewing the meaning of strength and exploring how ceramics behave differently from metals when we try to break them. Then, we can move on to talk about a couple of the new ceramics—for example, silicon nitride and silicon carbide—that have such high strength and stability. As we go, we'll discuss some of the new products in which these new ceramic materials are working their magic.

CERAMICS AND STRESS

How Stress Operates

A material breaks when a load (a *stress*) is applied to it. The stress actually tries to pull the atoms of the material apart from each other, first to form a crack and then to force the crack through the material until the material breaks into two or more pieces. Scientists think of the amount of stress needed to start a crack as *strength* and the amount of stress needed to drive the crack through the material as *toughness*.

Ceramics are very sensitive to stress. The atoms in a ceramic are bonded together to form a relatively rigid structure, similarly to the way metal girders in a building are interlaced and bolted or welded together. When a stress is applied to this type of structure, the structure doesn't deform easily. Instead, the ceramic takes the full force of the stress at the point where the stress is applied, such as when a baseball hits the living room window or a rock hits your car window. Even worse, the stress is magnified anywhere the ceramic has a flaw or notch. That's why a dish or a glass with a crack is so easy to break. That's also why you can easily trim a sheet of glass to the size you want by scratching it with a sharp tool and breaking it along that mark. The scratch is a notch that magnifies the stress, and the glass fractures along the scratch. Pores, inclusions, cracks, grooves from a grinding wheel, and even the grain boundaries in the microstructure are other flaws that can magnify a stress to break a ceramic material.

Most metals respond differently to a stress. Remember, from our discussion in the previous chapter, how metals are bonded by electrons that can move throughout the whole piece of metal? This bonding, plus the way the atoms are stacked in a metal, give the metal a special type of flexibility: The atoms can slide past each other when the metal is under stress. Such sliding allows the metal to bend slightly or *deform*. Called *ductility*, this deformation spreads the stress over a larger area so that it doesn't magnify and cause the metal to break.

Types of Stress

The particular way in which a stress is applied to a ceramic also is important to how the ceramic will react. As shown in Figure 5-1, a stress can be applied in several ways: by compression, bending, or tension. A stress that tries to smash, squash, or compress the ceramic material—for instance, that applied by a trash compactor—is called a *compressive* stress. Remember the scene in the movie *Star Wars*, where Luke Skywalker, Han

Solo, Princess Leia, Chewbacca, and C-3PO were trapped in the garbage bin and the walls were slowly closing in on them? If their robot friend R2-D2 hadn't shut down the compactor in time, they all would have been crushed by a compressive stress.

Compressive stress affects a ceramic by pressing against the defects in the ceramic and trying to make them smaller, so is not effective at breaking the ceramic. A ceramic can withstand a great compressive stress, or *load,* without fracturing or deforming, which is why ceramic brick and concrete are good materials for construction. Later in this chapter, we'll explore the use of the new ceramic material silicon nitride as balls for bearings. Silicon nitride ball bearings can withstand incredibly high compressive stresses.

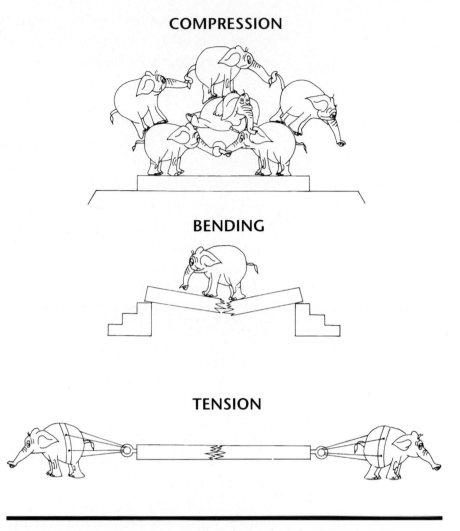

COMPRESSION

BENDING

TENSION

Figure 5-1. *Ceramics are strong in compression but weaker in bending and tension.* (Illustration drawn by Angie Egan, Salt Lake City, UT.)

Stresses caused by bending and tension also are easy to visualize. If you climb out on the limb of a tree, you're applying a *bending stress* to the limb. If you attach a rope to the limb and swing from it, you're subjecting the rope to tension, known as *tensile stress.* Bending and tension try to pull a material apart and to make any of the defects, or flaws, larger. This pulling apart magnifies the stress in a ceramic, just like the scratch does on the surface of a sheet of glass. As a result, most ceramics aren't very strong under a tensile stress.

Dealing with Stress

People have always been comfortable dealing with ceramics under compressive loads. The ancient Egyptian pyramids and many beautiful temples and churches were built with Nature's ceramics—stone. The foundations of our houses, buildings, and bridges all are constructed of ceramics. Engineers, however, have shied away from dealing with ceramics under bending and tensile stress—that is, until the last 30 to 40 years. During that time, new ceramics have been invented that have remarkably high strength and toughness, even under bending and tensile loads. Some of these new ceramics have resulted from the quest for materials that can withstand the high stress and high temperature inside jet engines and advanced automobile engines. Others have resulted from efforts to develop high-strength fibers to reinforce structures such as the wings and bodies of advanced lightweight aircraft. The next sections of this chapter tell about some of these remarkable new ceramic materials, the severe conditions they're able to survive, and the ways they're being used in new products that will benefit all of us.

SILICON NITRIDE

Silicon nitride, made up of atoms of silicon and nitrogen, is one of the special new ceramics invented by scientists. Silicon is one of the most abundant chemical elements in the Earth's crust, and nitrogen makes up nearly 80 percent of the air we breathe. For some reason, though, silicon and nitrogen never joined together when the Earth was being formed, so silicon nitride has never been found as a natural material on Earth. Scientists discovered silicon nitride when they heated a powder of silicon to a very high temperature (above 2200°F) in a closed ceramic box filled with nitrogen gas. They found that the silicon and nitrogen reacted together—actually combined together chemically—to form a ceramic with silicon atoms strongly bonded to nitrogen atoms. The scientists learned that this new ceramic substance was more resistant to stress, high temperatures, and rapid changes in temperature (*thermal shock*) than any other ceramic they had studied. Much of this early work on silicon nitride was conducted in England during the 1960s, when engineers were looking for ceramic materials that might survive inside gas turbine (jet) engines.

Silicon Nitride and Jet Engines

Gas turbine engines are extremely important in our modern world: They're the powerful engines that you see on the wings and tails on all of the jet airplanes lined up on airport runways or flying overhead to destinations around the world. In fact, many people think of a gas turbine engine as a jet engine. But gas turbine engines go far beyond aircraft propulsion. They're also an important and efficient means of generating electricity. They provide heat and power to run many industrial processes, such as chemical synthesis, papermaking, petroleum conversion to useful products, and even food processing.

Gas turbine engines come in all sizes, from those that can be carried by a single person and used to generate electrical power for a mobile medical unit, to those that are too big to fit into your living room and can generate enough electricity to run a whole hospital. Some of these gas turbine engines pump oil through the Alaska pipeline, 24 hours a day, through some of the most rugged countryside and worst weather on earth. Gas turbines were selected for that job because they're highly reliable. Other gas turbines produce steam to pump into oil wells to help recover more oil or provide the power for oceangoing ships.

> **USES FOR GAS TURBINE ENGINES**
>
> **Jet engines**
>
> **Electricity generation**
>
> **Emergency power for hospitals, computer systems, air-traffic control equipment**
>
> **Power and heat for industrial processes**
>
> **Oil pumping**
>
> **Power for seagoing ships**

Ceramics are of interest to the engineers who design gas turbine engines for good reason. Ceramics are lighter in weight than the metals that are currently used in turbines and have the potential to be used at much higher temperatures. The higher the temperature at which the materials in a gas turbine engine can operate, the more power is produced per gallon of fuel, and the less pollution is emitted into the air. Over the years, engineers have progressively increased the temperature inside gas turbines to the point that metals have now reached their upper temperature limits. Further gains in efficiency—less fuel consumption, lower pollution emissions, higher power output—require higher-temperature materials: ceramics.

Why can't we just make ceramics into the shape of the metal parts and build them into the engines in place of the metals? Engineers tried in the 1950s, but they weren't successful. The conditions inside a turbine engine are incredibly severe, and the ceramics available in the 1950s and 1960s weren't successful. Try to picture the conditions inside a turbine engine. When the engine is started up, the materials on the inside are heated from whatever the outdoor temperature is at the time (which could even be −60°F) to over 1800°F within a few seconds. That's like zapping an ice cube with a propane torch! Some special high-temperature metals, called superalloys, can survive this severe thermal shock because of their high strength, ductility, and high toughness; but most ceramics fracture with much less of a temperature change—even as small as from room temperature to around 400°F (like when you put a dish into a preheated oven).

Besides thermal shock, the materials inside a turbine engine must survive other severe conditions. Once the engine has started, the materials must perform acceptably above 1800°F for thousands of hours under a tensile stress that can range, in different regions of the engine, from 10,000 to 40,000 pounds per square inch (psi). That stress is about equal to lifting one to four adult elephants off the ground with a 1 inch diameter cable constructed from the material. To make matters even more difficult, the engine materials are continuously bathed in the high-temperature, erosive, and corrosive exhaust gases from the fuel burned in the engine. It's amazing any material can survive these conditions.

The first silicon nitride materials weren't strong enough to survive in a turbine engine: They only had a strength of about 10,000 psi. In the late 1960s, however, ceramic engineers in England attracted international attention when they developed silicon nitride that was 10 times stronger. Soon, numerous programs were launched to investigate the new wonder ceramic. I had just graduated from college and was fortunate to get to work on one of those programs. It was really exciting to participate in the development of a new family of materials!

Once international interest in silicon nitride had been generated, major ceramic-turbine development programs were started in the United States in the early 1970s. The following discussion uses one of the early programs to explain how a gas turbine engine works, where silicon nitride ceramics were used in the engine, and the benefits the engine designers hoped the ceramics would provide.

Figure 5-2. *Cutaway depiction of one type of turbine engine.* (Courtesy of AlliedSignal Engines, Phoenix, AZ.)

The program was sponsored by the Advanced Research Projects Agency of the U.S. Department of Defense and conducted by the AiResearch Manufacturing Company of Arizona (now AlliedSignal Engines). Figure 5-2 shows a cutaway view of the engine used in the program. It was a turboprop engine, a gas turbine engine that runs the propeller of an airplane. A gas turbine engine takes air from outside, compresses (pressurizes) the air and mixes it with fuel, burns the fuel–air mixture at high temperature in a *combustor*, and forces the hot air at high speed through a fanlike *rotor*. This process causes the rotor to spin, just like the wind turns a windmill, except that the rotor in the turbine typically spins at a much higher speed—usually more than 10,000 revolutions per minute (rpm). The rotor is mounted onto a shaft, so that when the rotor spins, the shaft rotates. In the turboprop engine, the shaft is connected to the gearbox of the propeller.

Look at Figure 5-2 and locate each of the major parts of the engine and identify those that were replaced by silicon nitride. Outside air is sucked into the engine through the large opening, or duct, at the upper left. The air moves to the right and is compressed by the two large, circular metal parts that look like the agitators in a washing machine, located about in the center of the engine. Fuel is mixed with the pressurized air after it leaves the compressor and is ignited to burn at flame temperatures of nearly 4000°F. This temperature is hotter than the metal engine parts can survive, so cool air is mixed in to reduce the temperature inside the engine to around 1800°F, still pretty hot. Now look just to the right of the compressor. See the three sets of fan blades? The hot gases pass through these fan blades. Each set consists of a row of blades that does not rotate—called the *nozzle guide vanes*—and the rotor, which does rotate. The nozzle guide vanes and the rotor blades are the parts that were replaced with silicon nitride.

When the nozzle guide vanes and the rotor blades were replaced with silicon nitride, the engine could run at 2200°F rather than the 1800°F that the metals were limited to. This increase in operating temperature increased the power output of the engine by 30 percent and decreased the fuel consumption by 7 percent. That early program demonstrated that silicon nitride could survive the severe conditions inside a gas turbine engine and also provide substantial benefits. However, the program also demonstrated that the silicon nitride materials available in the late 1970s weren't good enough for the long times (usually more than 10,000 hours) that turbines need to run in most applications.

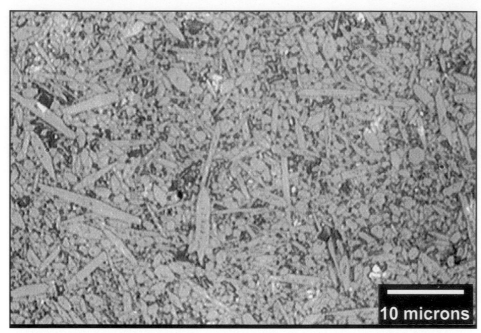

Figure 5-3. *Polished section of AlliedSignal grade AS-800 silicon nitride at high magnification, showing the unique microstructure. The highly fibrous, interlocked grains contribute to the very high strength and toughness.* (Photograph courtesy of University of Dayton Research Institute, Dayton, OH.)

Millions of dollars were spent worldwide from about 1980 to the present in trying to improve silicon nitride so that it would survive for longer than 10,000 hours in turbines. We now have silicon nitride materials with the remarkable strength of nearly 150,000 psi. How strong is 150,000 psi? A cable of this ceramic 1 inch in diameter—about as big around as a broom handle—could theoretically lift 50 automobiles, each weighing 3000 pounds. These same silicon nitride materials retain a strength of almost 100,000 psi at 2500°F and have been demonstrated to not break under a sustained stress of 15,000 psi at 2500°F for nearly 10,000 hours. For comparison, 1975 silicon nitride materials couldn't sustain a stress of 15,000 psi at 2500°F for even one hour.

In addition to increases in strength, improvement in the toughness (resistance to fracture) of silicon nitride has made it about three times more fracture-resistant than normal ceramics. This improvement in toughness was accomplished by forming a microstructure like that shown in Figure 5-3. The grains are highly elongated and intertwined, making it difficult for a crack to pass through, just the way the intertwined threads in cloth help the cloth resist tearing.

Despite the tremendous improvements in silicon nitride since 1975, trying to put it and other ceramic materials in turbine engines has definitely turned out to be one of the most difficult challenges ever faced with

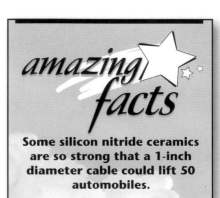

amazing *facts*

Some silicon nitride ceramics are so strong that a 1-inch diameter cable could lift 50 automobiles.

Figure 5-4. *Silicon nitride seal runner.*
(Photograph courtesy AlliedSignal Ceramic Components, Torrance, CA.)

ceramics. Finally, though, silicon nitride ceramics are entering into production for some gas turbine components. In 1996, AlliedSignal Engines introduced a silicon nitride *seal runner* into one of their turbine engines, a turbofan jet engine used to propel many types of business aircraft and some military aircraft. The seal runner is attached to the engine shaft and rotates up to 600 feet per second while it maintains contact with a stationary surface to seal in lubricating oil. This main shaft seal has always been a major reliability problem and cause for unscheduled repairs. The silicon nitride seal runner seems to have solved this long-standing problem.

Silicon nitride nozzle guide vanes are nearing commercial production in another AlliedSignal engine called an *auxiliary power unit* (APU). Every large passenger airplane has an APU (in addition to the propulsion engines) to provide air conditioning, heating, and other on-board power requirements. APUs also are in the airport ground carts that supply power for air conditioning while a plane is on the ground and to start the main engines. More than 100,000 hours of testing have been accumulated with silicon nitride vanes in engines installed both in ground carts and on aircraft under normal service. One individual engine ran for 7000 hours before it was removed and the ceramic vanes inspected. The silicon nitride vanes looked almost like new, compared with metal vanes that erode and corrode after only 2200 hours of engine operation.

Silicon nitride is being tested in larger gas turbine engines, too. Under funding by the U.S. Department of Energy, Solar Turbines of San

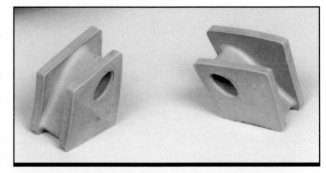

Figure 5-5. *Two silicon nitride nozzle guide vanes for an APU.*
(Photograph courtesy of AlliedSignal Ceramic Components, Torrance, CA.)

Figure 5-6. *Silicon nitride air turbine starter attached to a metal shaft.* *(Photograph courtesy of AlliedSignal Ceramic Components, Torrance, CA.)*

Figure 5-7. *Photograph of the Solar Turbines Centaur engine field–tested with silicon nitride rotor blades and a silicon carbide composite–combustion chamber liner.* (Courtesy of Solar Turbines, Inc., San Diego, CA.)

Figure 5-8. (left) *Gas turbine engine rotor with silicon nitride blades mounted in a metal disk. Magnified single blade shown below (right).* (Courtesy of Solar Turbines, Inc., San Diego, CA.)

Diego, CA, has designed, built, and field-tested a 5000 horsepower engine with silicon nitride turbine rotor blades. This engine accumulated about 1000 hours of operation during 1997 in an oil field in California, where it generated electrical power and also produced steam to be pumped into oil wells. The engine—with ceramic components—demonstrated a significant reduction in pollutants and an increase in power output.

Silicon Nitride for Cutting Metals

A later chapter will cover how ceramics are used at very high temperatures to extract metals from the rocks in which they are found and to form the resulting purified metals into shapes. In most cases, though, metal parts still are not in the final shape needed for a product such as an automobile engine. In fact, in some cases as much as one-half of the metal must still be cut (machined) away. This machining is done with various *cutting tools* that precisely drill, mill, cut, or grind away the unwanted metal. The more rapidly the metal can be removed, the more economically the part can be made, and the less we'll have to pay for the product.

Silicon nitride was discovered in the mid-1970s to be a super cutting-tool material able to machine cast iron and superalloys five to six times more rapidly than previous cutting-tool materials had been able to do. Such speed may not seem amazing to you, until you imagine the severe conditions the ceramic tool must survive to perform this feat. Imagine a small piece of ceramic about one-half inch square and slightly less than one-quarter inch thick (see Figure 5-9). Now imagine clamping the ceramic firmly into a metal holder (see Figure 5-10) and jamming the edge of the ceramic into a piece of metal that's moving

Figure 5-9. *Silicon nitride ceramic cutting tool inserts.* (Photograph courtesy of Ceradyne, Inc., Costa Mesa, CA.)

Figure 5-10. (left) *Ceramic cutting tool inserts and metal holders.* (Photograph courtesy of Kyocera.)

(rotating in a machine called a *lathe,* like a tire rotates on the axle of your car) more than 60 miles per hour. The edge of the ceramic gouges into the surface of the metal with enormous stress and begins to remove (machine away) a thin strip, or ribbon, of metal. The stress and friction involved causes the edge of the ceramic to heat up to around 2000°F almost instantaneously. Silicon nitride can survive this abusive treatment and successfully machine away about 300 cubic inches of metal per minute; that's equal to a metal block 10 inches wide, 10 inches long, and 3 inches thick. A similar metal-removal rate can be achieved by holding the metal piece stationary and moving the ceramic cutting tool, as in a drill or a milling machine. Furthermore, the movement of the cutting tool can be controlled by a computer to make very precise shapes.

Silicon nitride cutting tools have dramatically decreased the cost of machining cast iron parts for automobiles, trucks, and many types of industrial equipment, helping to keep the price down for many of the products we all buy. These cutting tools also have reduced the cost of precision-machining superalloy metal parts for the wide range of gas turbine engines discussed earlier.

Silicon Nitride for Bearings

One of the newest and most exciting uses of silicon nitride is for the balls in bearings. Bearings are required in just about everything that rolls or rotates. Do you have an old pair of roller skates? Look at one of the wheels at an angle from the side. You should be able to see a circular row of small metal balls, held in place on the outside and inside by metal rings. These metal parts are the bearings. A similar, but larger, bearing fits between

WIDELY VARIED USES OF SILICON NITRIDE BEARINGS

Skates, skateboards, and street luges

Racing bicycles

Racing cars

High-speed, hand-held grinders

Gas meters

High-speed compressors

Saltwater fishing reels

High-speed train motors

Photocopier rolls

Turbochargers

Figure 5-11. *Silicon nitride bearings in in-line skates boast record-setting performance and long life.* (Photograph courtesy of Saint-Gobain/Norton Advanced Ceramics, East Granby, CT.)

the axle of your car and each wheel. The mixer in your kitchen, the electric drill and bicycles in your garage, the in-line skates in your closet, and the swirlers in your hot tub all have ball bearings. All these bearings are traditionally made of special grades of steel.

Bearings let things move more easily by a rolling motion in place of a sliding motion. Each bearing ball, however, takes turns supporting most of the weight of the car or whatever the bearing is in. If you car tire hits a pothole in the road, the stress on the bearing is even greater. Since the load on a bearing is trying to crush the bearing ball, the stress in the ball is compressive.

Ceramic materials are hard, very resistant to wear, and strongest under compressive loading. These exact characteristics should make ceramics ideal for bearings. Ceramics such as synthetic ruby and sapphire were used for decades for bearings in watches and technical instruments, where the stresses were very low. However, when ceramic bearings were tried in new high-stress applications, such as high-speed motors and the main bearings for the shafts that support the rotors in turbine engines, they failed immediately. Even the best steel bearings didn't survive these difficult uses as long as desired by the equipment manufacturers.

USES FOR SILICON NITRIDE BEARINGS IN INDUSTRY
Food-processing equipment
Textile equipment
Chemical-processing equipment
Machine-tool spindles
Checkvalve balls
Pumps
Semiconductor processing equipment
Air-driven power tools

Then silicon nitride came along; it was the first ceramic tested successfully as a high-speed, high-load bearing. This feat was accomplished in about 1972, but those early ceramic bearings cost about one hundred times more than steel bearings. It took engineers until nearly 1990 to get the cost down enough to make silicon nitride bearings affordable. Even now, the cost of silicon nitride bearings ranges from two to five times more than that of metal bearings, but the benefits have proven so great that equipment manufacturers are willing to pay extra for the ceramic bearings. Between 15 and 20 million silicon nitride ball bearings were produced in 1996.

Why has silicon nitride become so popular as a bearing material? There are a number of reasons. Silicon nitride is three times harder (resistant to shape change or scratching under pressure) than bearing steel, can be surface-ground and polished to an extremely

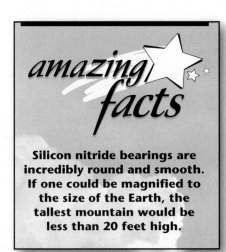

amazing facts

Silicon nitride bearings are incredibly round and smooth. If one could be magnified to the size of the Earth, the tallest mountain would be less than 20 feet high.

Figure 5-12. *Silicon nitride bearing balls and other bearing and wear-resistant components.* (Photograph courtesy of Saint-Gobain/Norton Advanced Ceramics, East Granby, CT.)

smooth surface so that it has 80 percent lower friction than steel, and wears at only one-seventh the rate of the best steel bearings. The extremely smooth surface is round and smooth to less than one-half millionth of an inch. If we could magnify one of these bearings to the size of the Earth, the tallest hill would be less than 20 feet high.

Another benefit of silicon nitride bearings is weight. A silicon nitride bearing is about 60 percent lighter than a steel bearing of the same size, which means that silicon nitride bearings use about 15 to 20 percent less energy to operate and can be run at up to 80 percent higher speed. Silicon nitride bearings also are smoother- and cooler-running than metal bearings and can even be operated at temperatures up to about 1000°F. All things considered, silicon nitride bearings last three to ten times longer than metal bearings.

The use that really launched silicon nitride bearings in the late 1980s was as *hybrid bearings* for high-speed machine tools. A hybrid bearing consists of ceramic balls or cylindrical rollers held in place by inner and outer metal rings (races). Hybrid bearings take advantage of the high compressive strength and light weight of the ceramic and the high tensile strength of the metal. Machine tools include the lathes, mills, drills, and grinding machines that the cutting tool inserts discussed earlier fit into, to cut and grind metals to their final shapes for products.

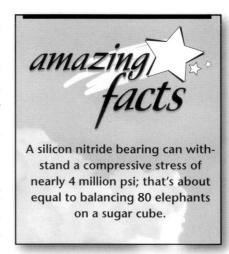

amazing facts

A silicon nitride bearing can withstand a compressive stress of nearly 4 million psi; that's about equal to balancing 80 elephants on a sugar cube.

Modern industry runs on machine tools, especially computer-controlled cutting and milling machines that shape the millions of metallic

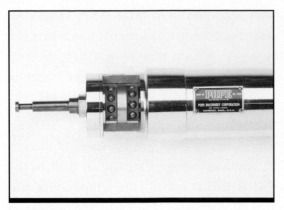

Figure 5-13. High-speed, rigid machine tool spindle with silicon nitride bearings. (Photo courtesy of the Pope Corporation, Haverhill, MA.)

parts required for automobiles, jet engines, appliances, and other machines. Silicon nitride bearings allow the machine tools to be constructed with higher stiffness, or rigidity. This greater rigidity decreases vibration, permitting the machine to run smoothly at higher

SILICON NITRIDE BEARINGS IN AEROSPACE
Gas turbine engines
Space Shuttle main engine
Space Shuttle liquid oxygen pumps
Helicopters
Aircraft anti-icing valves
Aircraft wing-flap ball screws
Gyroscopes
Military missiles

speed and to produce better parts at lower cost. Furthermore, the ceramic bearings last longer, resist corrosion (chemical attack) and wear, consume less energy, and can even be run with little or no lubrication. In contrast, if a metal bearing loses lubrication, it "freezes up" within seconds or minutes and fails. Silicon nitride bearings are the magic behind a whole new generation of machine tools.

Silicon nitride bearings are also superior for many other applications in industry, aerospace, medical technology, transportation, and even sports. They're the most durable and smoothest running bearings you can

DENTAL HANDPIECE COMPONENTS

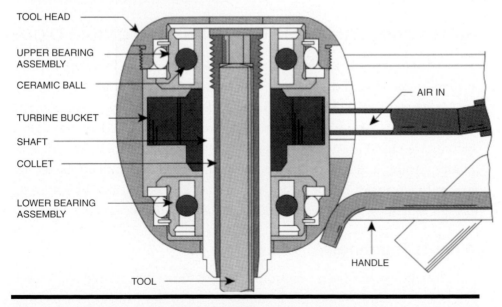

TOOL HEAD

UPPER BEARING ASSEMBLY

CERAMIC BALL

TURBINE BUCKET

SHAFT

COLLET

LOWER BEARING ASSEMBLY

AIR IN

HANDLE

TOOL

Figure 5-14. Schematic of a dental handpiece, showing ceramic bearings. (Courtesy of Den-tal-ez Inc., Lancaster, PA.)

get for your in-line skates and mountain bikes. They're used in race cars and will be used in passenger cars and trucks when the price comes down a bit more. Silicon nitride bearings are even being evaluated for computer disk drives, which could require over 60 million balls per year.

A particularly interesting and demanding use for silicon nitride bearings is for the butterfly valves in commercial aircraft such as the Boeing 777. These valves regulate the air flow in the environmental control system that adjusts the temperature and pressure of the air inside the airplane. The bearings operate at temperatures as high as 700 to 900°F and under a stress of 600,000 psi. Federal Aviation Association standards require the valves and bearings to last at least 625,000 cycles.

Another interesting use for bearings is in dental drills, which operate at the incredible speed of 450,000 rpm. Dental instruments must be thoroughly sterilized after each patient to protect against diseases. This sterilization process has become especially important to protect against AIDS. The dentists are required to use an *autoclave,* which is similar to the pressure cooker we sometimes use in our kitchen to quickly cook a roast or stew. The autoclave produces high pressure steam at several hundred degrees Fahrenheit. Those conditions kill any bacteria and viruses, but also damage the lubricant for the bearings in the dental drill. Because metal bearings can't work without effective lubrication, they don't hold up to the autoclaving procedure. Silicon nitride bearings, which can operate with little lubrication, solve the problem.

HIGH-STRENGTH, HIGH-TOUGHNESS ZIRCONIUM OXIDE

Another remarkable ceramic developed since 1970 is known as *transformation-toughened zirconia (TTZ).* Zirconia is a shorthand way of saying *zirconium oxide,* which is a ceramic made up of atoms of zirconium and oxygen. Transformation-toughened zirconia is related to the cubic zirconia we talked about in Chapter 3, but it has been specially modified by adding yttrium, calcium, or magnesium atoms so that it's much tougher—in fact, so tough that it can be hit with a hammer and not break. The reason for its high toughness seem magical and unique for a ceramic, as you'll see soon.

What Is Toughness?

Toughness has been mentioned several times but hasn't really been described. To understand the meaning of toughness, imagine a piece of material containing a small crack. When you apply a tensile stress, the crack will get bigger and travel (propagate) through the material. If the material has low toughness, the crack will propagate easily and quickly, and the material will break. If the material has high toughness, the crack won't propagate easily, requiring a much higher level of stress to cause fracture.

Most ceramics have very low toughness; they have no natural resistance (such as the ductility in a metal) to keep a crack from slicing through their microstructure like a hot knife cutting through butter. The secret to increasing toughness in ceramics is to build little obstacles, or roadblocks, into the microstructure to interfere with the movement of a crack through the ceramic. For example, the intertwined fibrous grains in silicon nitride, shown earlier in Figure 5-3, force the path of a crack to twist and turn. This twisting and turning is enough to make silicon nitride two to three times tougher than a typical polycrystalline ceramic and five to six times tougher than typical glass. Transformation-toughened zirconia is even tougher than silicon nitride.

Transformation Toughening

Transformation toughening also involves interaction of the microstructure with a crack, but it goes a step farther than silicon nitride. Rather than just sitting there and causing the crack to follow a tortuous path, the TTZ microstructure actually changes shape near the tip of the crack and stops the crack dead in its path. That's what's so magical about TTZ: It's the first ceramic with a built-in ability to stop a crack. Let's take a closer look at how TTZ stops a crack.

Figure 5-15. *Transmission electron microscope image showing the microstructure of transformation-toughened zirconia. These crystallites within a grain are only 100 to 400 nanometers long.* (Photograph courtesy of A.H. Heuer, Case Western Reserve University, Cleveland, OH.)

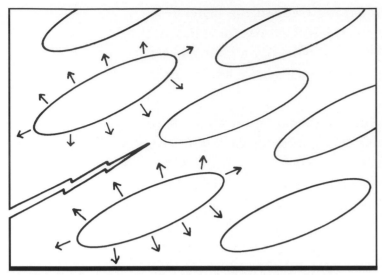

Figure 5-16. *Schematic illustrating the mechanism of transformation toughening. The crack allows zirconia crystallites to transform, which results in about 3 percent volume increase that wedges against the crack and stops it.*
(Illustration drawn by Angie Egan, Salt Lake City, UT.)

Transformation means *change.* A transformation in a ceramic involves an abrupt change in the arrangement of the atoms, which results in an abrupt change in the size of each grain in the microstructure. Such a change doesn't just happen on its own; it needs some form of stimulation, such as a force or a change in temperature. In the case of TTZ, the stimulus is a stress applied to it and the magnified stress that results at the tip of a crack. As discussed earlier, this stress magnification usually causes a ceramic to break, but not TTZ. Instead, the TTZ is stimulated to transform, and the transformation causes tiny crystallites in each grain around the tip of the crack to expand, getting slightly bigger. This expansion presses against the crack, placing it in compression, and keeps it from propagating.

Figure 5-15 shows the microstructure of one form of TTZ, and Figure 5-16 shows how the tiny crystallites grains press against a crack. You can try a couple of simple experiments to get a feeling of how this works. The easiest experiment is to take a piece of string or thread, tie one end to a pencil, and hold the other end between your thumb and finger. If you hold the string very loosely and pull with the pencil, the string slides easily between your finger and thumb, like a crack going through a normal ceramic. Now press the string tightly with your finger and thumb and try to pull the end of the string attached to the pencil. Movement is much more difficult, like for a crack trying to go through TTZ.

A more elaborate experiment is to use a piece of wood about one-half inch by one-half inch and four or five inches long. Start a crack in one end with a chisel. Now try to wedge the crack open with the chisel or pull the crack open with your hands. The crack easily propagates until the wood has split into two pieces. Take a second piece of wood and tightly clamp a C clamp (available at all building and hardware stores) about two inches from the end. Now start a crack with your chisel and try to split the wood. It's difficult to get the crack to go past the compressive stress applied by the C clamp. This "roadblock" also is analogous—in an exaggerated way—to what's happening inside the TTZ ceramic.

Obviously, there's a lot more science behind the magic of transformation toughening than we've discussed, but it's more complex than we need to cover here. The key message is that TTZ is very special and unique; but there's still a little more to the story that you might find interesting. Finely ground up powder made from TTZ can be added to another ceramic so that it's dispersed throughout the microstructure of that ceramic. These tiny grains will transform when a crack tries to pass through, just like they do in pure TTZ. The best example of this is the addition of TTZ particles to alumina. Adding as little as 15 to 20 percent TTZ more than doubles the toughness of alumina. This new *transformation-toughened alumina (TTA)* is nearly as strong and tough as TTZ.

Uses for Transformation-Toughened Ceramics

Transformation toughening was a breakthrough in achieving high-strength, high-toughness ceramic materials that could be as durable and nonbreakable as metals. In fact, the mechanism is similar to toughening in some forms of steel, so that TTZ has sometimes even been called "ceramic steel." Now a whole family of transformation-toughened ceramics is allowing ceramics to move into many challenging uses that previously could be satisfied only by special, expensive metals or metal–ceramic composites (we discuss composites later in this chapter).

One of the earliest applications of transformation-toughened alumina was for cutting tool inserts similar to those described for silicon nitride. Silicon nitride inserts are good for cutting cast iron and superalloys but not for steel and some other metal alloys. Fortunately, TTA works well for these metals that silicon nitride doesn't work for.

Engineers have come up with a broad range of uses for TTZ, from watchbands to the tooling for making aluminum beverage cans. Remember the picture of the Rado watch in Chapter 1? The links for the wristband and the bezel (the border around the transparent sapphire watch glass) are made from TTZ. You can imagine how tough the ceramic has to be to not break, scratch, or chip under the daily abuse that a typical wristwatch endures. Ceramic engineers also have learned to make this TTZ for watches in about 10 different designer colors.

Other uses for "ceramic steel" also were mentioned in Chapter 1: scissors, knife blades, golf putters, golf cleats, and even indestructable buttons. Because the TTZ doesn't deform or wear as easily as a metal, the scissors and knife blades stay much sharper for much longer, and the knife blade

> **SURPRISING USES OF TRANSFORMATION TOUGHENED CERAMICS**
>
> Scissors
>
> Knife blades
>
> Buttons
>
> Cutting-tool inserts
>
> Papermaking machine parts
>
> Pump parts
>
> Wire-making tools
>
> Extrusion dies
>
> Golf clubs and cleats
>
> Hip replacements
>
> Aluminum-can forming tools

never passes on a metallic taste to food. TTZ golf cleats are the flattened type that look like they have little hills and valleys rather than long, sharp spikes. They give good traction but don't chew up the greens and don't wear down when you walk on cart paths. TTZ has even been built into the face of the driver, the golf club designed to hit the golf ball the longest distance. You've probably seen John Daly on television, winding up and smacking a ball over 300 yards. Can you imagine what would happen if he swung a golf club with that much force at a glass window or a ceramic vase? You'd have quite a mess to clean up. The TTZ built into the face of the golf club, though, can withstand this high level of stress over and over.

Many other objects now are made of transformation-toughened ceramics and are discussed in later chapters. Some examples include critical parts in the papermaking process, wear-resistant and corrosion-resistant parts in the chemical processing industry, tooling for making metal wire, and even artificial hips. We can now take advantage of some of the special characteristics of ceramics, such as resistance to wear and to chemical attack, without worrying so much about the ceramic breaking like a glass window or a ceramic vase.

HIGH-STRENGTH CERAMIC FIBERS

Ceramic engineers are always trying to make glass and polycrystalline ceramics stronger. They've learned that they can fabricate these materials into long fibers around 10 times stronger than the bulk glass or ceramics we've been talking about so far. These fibers aren't even as big around as a hair; they're similar in size and appearance to the filaments in thread. Take a close look at a piece of thread with a magnifying lens or a microscope. Each strand of thread is made up of hundreds of smaller filaments.

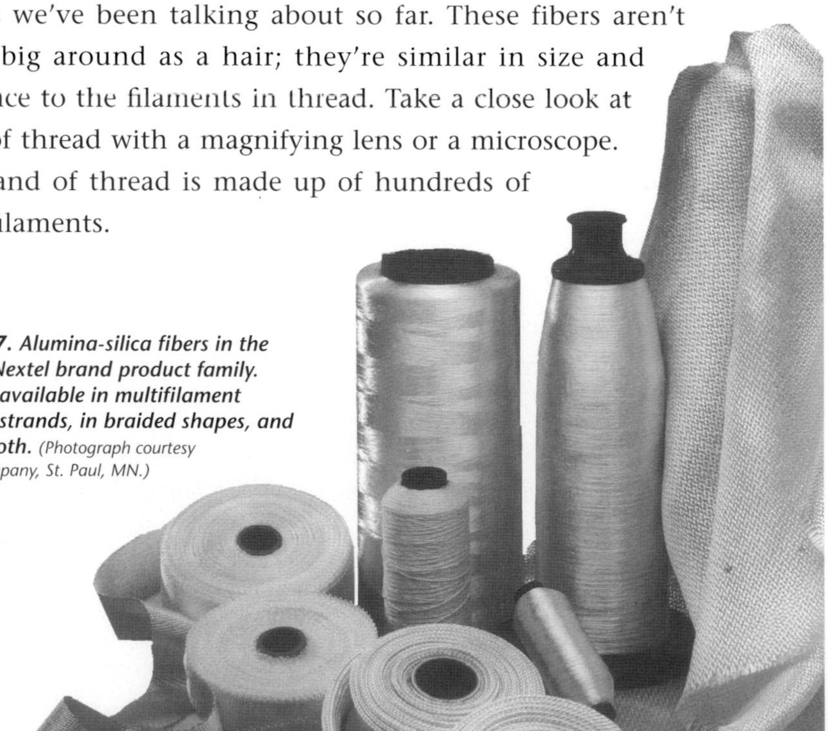

Figure 5-17. *Alumina-silica fibers in the 3M brand Nextel brand product family. The fiber is available in multifilament continuous strands, in braided shapes, and in woven cloth.* (Photograph courtesy of the 3M Company, St. Paul, MN.)

Figure 5-18. *Silicon carbide and glass fibers: woven multifilament silicon carbide fibers, large diameter single-filament silicon carbide fibers, and multifilament glass fibers. Sewing pin shown for size comparison.* (Photograph by D. Richerson.)

Why Are Glass and Ceramic Fibers So Strong?

Earlier in this chapter, we discussed the factors that determine the strength of a ceramic material, especially the effects of defects such as pores, inclusions, and cracks. The thin fibers can be fabricated with much smaller defects than those trapped when a bulk ceramic is made, so that the fibers are much stronger. Even higher strength has been achieved with tiny *whiskers* of single crystal ceramics such as silicon carbide, alumina and graphite. These whiskers range from about 4 to 24 millionths of an inch (100 to 600 nanometers) in diameter, less than one-tenth the thickness of a hair, and are nearly 20 times stronger than a bulk ceramic. They have been added to metals and ceramics as reinforcement, but they are relatively expensive and also pose a health hazard, similar to asbestos, during fabrication.

The Use of Ceramic Fibers in Composites

The long fibers (usually referred to as *continuous fibers*) have proved to be cheaper than whiskers, as well as nonhazardous, and they've become important for many products.

The earliest long fibers were made by flowing molten glass through a small hole and pulling it into a long strand. You've probably been to a craft fair where you watched a glass artist at work and saw how easy it is to stretch the hot glass into a thin strand. Engineers learned, in the 1930s, that they could make many strands at one time by letting the molten glass flow through a plate or block with many holes. In fact, if they blew air at high speed under the plate or block, fibers would be pulled out and formed into a wooly bundle that made great house insulation. This product is the fiberglass insulation that we talked about in Chapter 1, which has saved enough energy since about 1938 to provide all of the energy requirements for the United States for a year (remember the staggering 25,000,000,000,000,000 Btu?).

Engineers also learned to make clusters of glass filaments into long strands of thread that could be woven like cloth. They found that they could bind the fibers together with a plastic (a polymer) such as polyester or epoxy to form a solid engineered material called a *composite*. The plastic in a composite forms a *matrix* that holds together and protects the fibers; the fibers provide high strength and stiffness (rigidity).

A composite combines the best characteristics of each material to make a new engineered material that's better than either material alone. You can demonstrate this by making your own composite. Thoroughly soak a long piece of string in glue, press the string and glue into a paper cup until the glue fills all of the spaces between the strands of string, and then let the glue harden. Your composite is a lot stronger and stiffer (less flexible) than the string and glue separately; but it's nowhere near as strong and stiff as a plastic material reinforced with glass fibers.

The first polymer matrix composites were made over 50 years ago with glass fibers and have become commonly known as *fiber-glass composites* or simply *fiberglass*. One of the most common uses of fiberglass is in boats. A form or mold in the shape of the boat hull is first made with wood or other material and is faced upwards, as if the boat were upside down. Strips of cloth woven from glass fibers are then coated with liquid plastic resin (which

---

ENGINEERED CERAMIC-REINFORCED COMPOSITES FOR CONSTRUCTION

Flooring

Roof shingles

Reinforced concrete

Bridge and pier decking

I beams

Pipe

Storage tanks

Utility poles

RECREATIONAL USES FOR CERAMIC-REINFORCED PLASTICS

Pleasure boats

Skis

Camper shells

Skateboards

Surfboards

Kayak paddles

Hot tubs

Fishing poles

Tennis rackets

can be purchased at a hardware store) and layered over the mold. This process is repeated until the top surface of the mold is completely covered with the desired number of layers. The resin is allowed to cure (harden). The hardened, fiber-reinforced shell forms the shape of the hull of the boat and can then be lifted off the mold.

Aircraft structures, camper shells, hot tubs, cafeteria trays, various containers, skateboards, surfboards, and many other products have been constructed by the *glass–cloth lay-up technique*. Other products, such as fishing rods, have been constructed with the fibers all aligned in one direction. Many other parts are formed with chopped (discontinuous) fibers mixed with a resin and shaped by pressing them into a shaped cavity (like our paper cup) rolling or spraying them into a sheet, or squeezing them through a nozzle (a process called *extrusion*) into a rod or tube. Many automobile parts, which need to be lightweight and low cost but do not require maximum properties, are made using these discontinuous-fiber processes.

Ceramic fibers in a polymer matrix also are becoming important for construction. A bridge 328 feet long and 184 feet wide was recently constructed completely from glass-reinforced polyester. The polymer-matrix composite has much better corrosion resistance than metals and is projected to greatly reduce maintenance.

Polymer-matrix composites reinforced with ceramic fibers have some of the highest strengths of any materials in proportion to their weight. Thin, lightweight structures made of these composites can often perform a task that previously would have required a heavy metal structure. The fibers with the highest strength and stiffness per unit of weight are made of carbon/graphite, a ceramic composed completely of the lightweight element carbon.

Composites with High-Strength Carbon Fibers

amazing facts

A Boeing 757 Aircraft contains more than 3000 pounds of carbon–epoxy composite. A B-2 stealth bomber is estimated to contain about 50,000 pounds of carbon–expoxy composite.

Carbon fibers in epoxy and other advanced polymers were introduced in the early 1970s and have become important composites for aircraft, as a replacement for aluminum metal. For example, the V-22 vertical takeoff aircraft—which has propellers that are vertical during takeoff and landing but horizontal during cruise—is constructed with over 6800 pounds of a

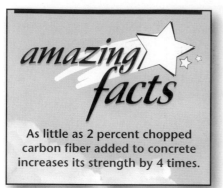

composite made up of carbon fibers in an epoxy matrix; that's about one-half the structural weight of the aircraft. Over 10 percent of the structural weight of the F-18 fighter jet is carbon–epoxy, and the B-2 stealth bomber is estimated to contain between 40,000 and 55,000 pounds of carbon–epoxy composite.

Lightweight composites also are important for commercial aircraft. The Boeing 757 uses about 3340 pounds of carbon–epoxy composite, which is a weight savings of about 1490 pounds. This decreased weight saves a tremendous amount of fuel. Other commercial aircraft such as the DC-10, L-1011, 737, 767, and 777 all contain substantial amounts of carbon–epoxy composites.

Carbon fibers are used in many more common ways, especially for sports equipment such as golf club shafts, kayak paddles, tennis rackets, and skis. They're also used, particularly in Japan, in chopped form as an additive to concrete for the construction of buildings. A 2 percent addition of carbon fibers to concrete quadruples its strength and increases the amount of deflection the concrete can tolerate without fracturing by nearly 10 times. This new reinforced concrete is much more able to withstand an earthquake than standard concrete, a quality important in Japan, where earthquakes are so prevalent.

Perhaps the most unusual use of carbon-reinforced epoxy is for treating ingrown toenails. A small thin strip of the composite is bent and glued to the problem toenail. The carbon fibers are so stiff that they try to straighten the composite, which applies a steady pull to the top of the toenail, preventing it from growing inward into the toe.

USES FOR HIGH–TEMPERATURE CERAMIC–MATRIX COMPOSITES

Rocket nozzles

Radiant burners

Combustor liners

Space Shuttle nose and wing leading edges

Tubes to salvage waste heat

High-temperature filters

Fire walls

Brake linings for airplanes

High–Temperature Ceramic–Matrix Composites

The polymer-matrix composites we've discussed so far are great at room temperature but lose their strength when the temperature is increased above a few hundred degrees Celsius, and actually melt or burn at still higher temperatures. Researchers are presently developing composites with ceramic fibers in a ceramic matrix. These ceramic-matrix composites (CMCs) can be used at much higher temperatures than polymer matrix composites and even metals withstand. For example, a composite constructed with silicon carbide fibers in a silicon carbide matrix has been run as the

Figure 5-19. *Carbon fibers protruding from a fractured ceramic-matrix composite. The individual fibers are about three ten-thousandths of an inch in diameter.* (Scanning electron microscope image courtesy of D. Richerson.)

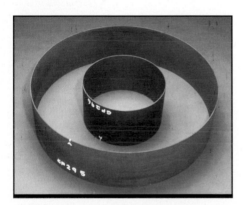

Figure 5-20. *Large, gas turbine engine ceramic-composite combustor liners fabricated from silicon carbide fibers bonded together with a matrix of silicon carbide from AlliedSignal Composites, Newark, DE. The larger liner is about 30 inches in diameter.* (Photograph courtesy of Solar Turbines, Inc., San Diego, CA.)

combustor liner of a gas turbine engine above 1800°F. The previous metal combustor liner required a lot of air cooling, which reduced the efficiency of the engine and also caused pollution. The CMC liner increases engine efficiency and lowers the pollution discharged into our air.

Although CMC technology is pretty new, some interesting uses have been found for CMCs. One company has made a porous, gas-fired heater for softening glass to shape car windows and for rapidly drying paint in automobile-production plants. This heater, called a *radiant burner*, reduces the usual fuel requirement by one-third. The same type of burner soon may be used for drying paper in papermaking machines and even for cooking hamburgers in fast-food restaurants.

Another company has developed a porous CMC material that can filter out the ash particles when coal is burned to produce electricity (which is the major method of generating electricity in the United States). Twenty years ago, all of the ash from coal burning went up the smokestack and added to air pollution. Technology was developed to filter out much of this coal ash, but the hot gases first had to be cooled to a temperature that metals could handle. This cooling process wasted a tremendous amount of heat energy that could have been better used to generate electricity. The new CMC filters work at a much higher temperature, so that the heat can now be used to produce electricity. These ceramic *hot-gas filters* will significantly reduce the consumption of natural resources and also the emission of pollutants.

The highest-temperature composites consist of carbon fibers in a carbon matrix and are known as *carbon–carbon composites*. Able to survive the high temperatures in rocket nozzles and on the nose and wing leading edges of the Space Shuttle during re-entry, these composites also are effective as brake linings for aircraft and cars.

CERAMIC STABILITY

Stability vs. Temperature

Another special characteristic of some ceramics is *stability*, especially at high temperatures. Organic materials burn, melt, or decompose at temperatures usually well below 750°F. Aluminum metal melts at 1220°F. Iron-based alloys (such as cast iron and steel) melt at a much higher temperature but become very weak above 1500°F and also react with oxygen in the air. Superalloys based on cobalt and nickel, and used for jet engines, have higher temperature capability but begin to lose strength above 1800 or 1900°F. Some silicon carbide ceramics can survive for thousands of hours above 2900°F. Zirconium oxide has even been used above 3700°F. Some of the many ways in which high temperature stability are important are discussed in Chapter 10.

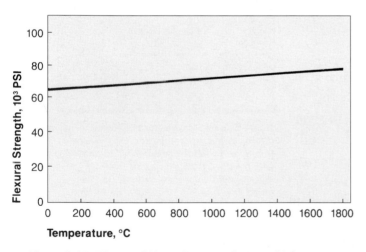

Figure 5-21. *Silicon carbide retains strength to very high temperature; strength measured in the bending mode.* (Courtesy of Saint-Gobain/Carborundum Structural Ceramics, Niagara Falls, NY.)

Dimensional Stability and Low-Thermal-Expansion Ceramics

A second special characteristic of some ceramics is dimensional stability during changes in temperature. Heat causes the atoms in a material to vibrate and move apart. Referred to by scientists as *thermal expansion*, this atomic movement usually results in growth or expansion of the material during heating, and contraction or shrinkage during cooling. Organic materials, nearly all metals, and most ceramics expand and contract a lot during heating and cooling. A very few ceramic materials—some that are glass and some polycrystalline—exhibit nearly zero thermal expansion.

Several problems are caused by a large amount of thermal expansion in a material. One concern we already mentioned is thermal shock. Materials, especially ceramics, with high thermal expansion are much more likely to break or be damaged by the rapid temperature change that occurs during thermal shock. Another concern is simply the changes in shape and dimensions that a material experiences during a temperature change. Let's look at how one low-thermal-expansion ceramic material, Pyrex glass, was used to remedy each of these concerns for two vastly different applications: ovenware and telescopes.

Have you ever taken a plate or other piece of glass or ceramic out of a hot oven and watched it break when you put it down on the stove top or counter top? Pyrex is a special low-thermal-expansion glass material that was developed so it could be taken safely out of a hot oven, or even taken straight out of the refrigerator and placed into a preheated oven, without breakage. But what does this quality have to do with telescopes? It's a fascinating story that starts back in the 1600s.

The first telescope was made in 1608 by the Dutch scientist Hans Lippershey, who mounted two transparent glass lenses in a hollow tube. The next year, Galileo Galilei constructed a larger telescope and began to explore the heavens. His small telescope allowed him to observe mountains

Figure 5-22. *200 inch mirror blank ready for shipping by rail from Corning, NY, to Pasadena, CA.*
(Photo courtesy of the Corning Incorporated Department of Archives and Records Management, Corning, NY.)

and craters on the moon, the moons of Jupiter, and 10 times more stars than were visible to the naked eye. Both of these early telescopes were referred to as *refractive telescopes*. Light passed through their transparent glass lenses and was bent, or refracted, as we discussed in Chapter 4. Controlling the shape and curvature of the lenses caused the light coming through the telescope to refract, giving the user a magnified image of the object being observed. Binoculars and telephoto lenses for cameras work in the same way.

Early astronomers learned that larger diameter lenses gave higher magnification. In 1668, Sir Isaac Newton found that a telescope could be made with mirrors instead of lenses and that large mirrors were easier to make than large glass lenses. His new type of telescope was called a *reflecting telescope*. Large reflecting telescopes with metal mirrors were built in the 1700s and 1800s; one such telescope was 40 feet long and 4 feet in diameter. These telescopes had a major problem, though. Variations in temperature caused their high-thermal-expansion metal mirrors to distort and blur the image of distant objects.

THE PALOMAR 200 INCH TELESCOPE MIRROR: A WONDER OF PATIENCE AND PRECISION

Required 10 hours to cast from 42,000 pounds of 2700°F glass

Was cooled very slowly during 10 months, to prevent cracking

Had more than 10,000 pounds ground away, using over 68,000 pounds of ceramic abrasives and jeweler's polishing rouge

Was polished to a precision of one-thousandth the thickness of a sheet of paper

Took over 15 years from start to finish

In the early 1900s, George Ellery Hale proposed making large mirrors from glass with just a very thin coating of metal for the reflecting surface. He believed that glass would be a better construction material for the mirrors, because its lower thermal expansion would assure better dimensional stability during temperature changes. However, large glass blanks had never been fabricated. Hale worked with a glass company to make first a 60 inch mirror and then a 100 inch mirror. The resulting telescopes, especially under noted astronomer Edwin Hubble, led to spectacular advances in our knowledge of the universe. Unfortunately, though, the glass mirrors still suffered enough thermal expansion to cause blurring.

By 1927, Hale had heard about a glass called *fused quartz*. Quartz, a natural ceramic made up of silicon and oxygen, is one of the most abundant materials on Earth, found in the form of sand and in the rocks sandstone and quarzite. With difficulty, quartz can be melted at high temperature to form a glass that has very low (near-zero) thermal expansion. Hale began a project to construct a 200 inch diameter reflecting telescope using fused quartz glass. After four years of effort, however, he gave up and switched to Pyrex glass, which had been developed by Corning Glass Works for ovenware. Pyrex glass did not have zero expansion, but it did have much lower expansion than the glass that had been used previously for mirrors.

Now the story gets interesting. Corning had not previously made a disk larger than 30 inches, so the task was pursued in a series of scale-ups from 30 to 60 to 120 inches, with engineering problems solved as they were encountered. Finally, after several more years of effort, the Corning team was ready in March of 1934 to cast (pour molten glass into a shaped mold, like pouring gelatin into a bowl) a 200 inch mirror blank.

The casting was accomplished by melting a master batch of 65 tons of glass at 2700°F; transferring the molten glass to ceramic-lined buckets (ladles), each holding 750 pounds of molten glass; and pouring the glass from each ladle into the ceramic mold for the 200 inch mirror blank. The mold, which looked a little like the bottom of an enormous waffle iron, took 10 hours to fill with 42,000 pounds of white-hot glass. The whole mold, filled with molten glass, was moved into a special oven, where it could be cooled very slowly to room temperature. Cooling was finally completed by June of 1934.

Figure 5-23. *Picture of a spiral galaxy in the constellation Pisces, taken with the 200 inch Hale Telescope.* (Photo courtesy of Hansen Planetarium, Salt Lake City, UT).

After all this time and effort, the cast glass disk contained defects that made it unsuitable for a telescope mirror.

By December of 1934, the problems causing the defects in the first 200 inch disk were understood and thought to be solved. Also by this time, an improved glass composition, with 25 percent lower thermal expansion, had been developed. A second 200 inch disk was cast with the new composition and carefully cooled for 10 months. This blank was declared suitable for grinding and polishing into a mirror. With substantial international fanfare, the enormous mirror blank was transported to the grinding shop in Pasadena, CA, on a special rail car. Grinding of the disk began immediately and was expected to be completed by the early 1940s, but work was interrupted by the Second World War.

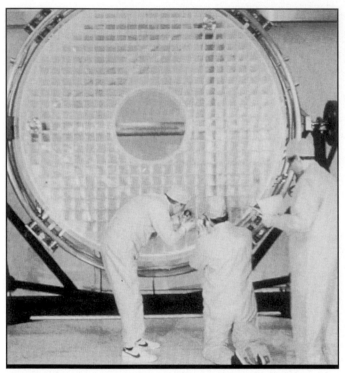

Figure 5-24. Glass mirror for the Hubble Space Telescope, showing the lightweight cellular construction. (Photo courtesy of Corning, Inc., Corning, NY.)

Grinding and polishing resumed after the war and were completed in 1948. Over 10,000 pounds of the mirror blank had been painstakingly ground away, consuming over 68,000 pounds of ceramic abrasives and jeweler's polishing rouge. The entire surface of the 200 inch disk was polished to a precision of four millionths of an inch, which is about one-thousandth the thickness of a sheet of paper.

More than 20 years after Hale started the project, the 200 inch telescope entered service, in January 1949, on Mt. Palomar near San Diego, California. Sadly, Hale had died in 1938, but the telescope bears his name and allowed us to see more deeply into space and more clearly than ever before. His idea of using a low thermal expansion glass was a great success.

The use of low-expansion, dimensionally stable glass didn't end with the Palomar Hale Telescope. The Hubble Space Telescope, which has a 94 inch low-thermal-expansion glass mirror, has allowed us to explore the heavens without atmospheric distortion or light pollution. This Earth-orbiting telescope provides images 10 times sharper than ever before and can detect starlight 50 times more faint than that detectable by earth-bound telescopes, extending our view even deeper into space. Able to detect the light from a two-battery flashlight on the moon, the Hubble Telescope is one of the most remarkable scientific and engineering feats of all time.

Figure 5-25. Low-expansion glass mirror for the 325 inch diameter Subaru Telescope. *(Photo courtesy of Corning, Inc., Corning, NY.)*

A whole new mirror technology had to be developed for the Hubble Telescope; the mirror had to be lightweight enough to be carried into orbit by the Space Shuttle and stable enough to withstand the rapid, large temperature changes that occur every time the telescope passes out of the Earth's shadow into the sunlight. Corning invented a new glass, called ULE (for *ultra-low expansion*), which changes dimensions less than 60 parts per billion per degree Celsius from 5° to 35°C; this means that a 100 inch diameter mirror would only change dimensions about three ten-thousandths of an inch over a 50-degree Celsius temperature change. To avoid casting a solid mirror blank, and to minimize weight, Corning learned to fuse together small segments of ULE glass to form a lightweight, cellular—or honeycomb—core with only a thin mirror surface. The Hubble mirror was constructed from 1800 pieces of ULE glass and weighed only 1650 pounds, 75 percent less than a solid mirror of the same size.

The technology acquired during the making of the Hubble Telescope has been applied to making larger ground-based telescopes. The Subaru Telescope, with a 325 inch diameter mirror made by welding together hexagonal ULE segments, entered into service on top of Mauna Kea, Hawaii, in 1998.

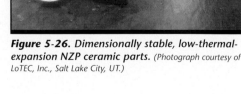

Figure 5-26. Dimensionally stable, low-thermal-expansion NZP ceramic parts. *(Photograph courtesy of LoTEC, Inc., Salt Lake City, UT.)*

Other Low-Thermal-Expansion Ceramics

Other important low-expansion ceramics or families of ceramics have been developed in recent years. One such ceramic probably familiar to you is Corning Ware, developed by Corning Inc.; examples that you probably haven't heard of are *cordierite* and NZP. Cordierite is made up of atoms of magnesium, aluminum, silicon, and oxygen and is used for the catalytic converters in automotive pollution control systems. This use is discussed in Chapter 9. NZP stands for *sodium zirconium phosphate* but is actually a whole family of ceramics, some of which have low thermal expansion. One form of NZP, for example, can be heated to over 2300°F and dunked into ice water without breaking.

OVERVIEW: A NEW DIMENSION OF STRENGTH AND STABILITY

Strength and stability present yet another magical dimension of ceramics. Not always the weak and fragile materials that we must handle with care, ceramics can also be some of the strongest, most durable materials. The remarkable advances of the past 30 years are just a start. Strong and durable ceramics and composites, with amazing new properties and applications, will evolve during the next 30 years and beyond. These products will be important in supersonic jet aircraft that can take us to the edge of space or around the world in just a few hours. They'll provide dimensionally stable building materials for the space station, high temperature linings for efficient new systems of electricity generation, and lightweight aircraft and engines of the future.

Ceramics and The Electronics Age

We clearly are living in the Electronics Age. Just think about it for a moment. If you were born in 1980, you've grown up with your hands on the keyboard of a computer or the joy stick of video games. When you were born, though, there were no desktop computers; and, until you were at least 10 years old, there was no Internet to surf. Your parents probably never saw a computer until well after you were born, which explains why you're always having to show them how to use one. Your grandparents didn't have hand-held calculators, even if they were mathematicians, scientists, or engineers. Their "calculators" were slide rules, which probably sound as ancient as abacuses—that's if you've even heard of either one before.

Electronics have taken over our lives in a remarkably short time. Every product we buy now seems to have electronics inside: our breadmaker that can be programmed to automatically make and bake a loaf of bread; our new clock that chirps each hour to the tune of a different bird; our microwave oven that's often better at judging how long to thaw or cook food than we are; and, of course, our desktop computers. The market for products run by electronics has been estimated at more than $15 trillion per year.

How great is the role of ceramics in electronics? Enormous! In fact, to the surprise of most people, the largest market for ceramics is in electronics. This chapter introduces you to the many roles that

PRODUCTS AND USES

Computer chips

CD players

Satellite communications systems

Electronic clocks

Automobile electronic ignition systems

Programmable appliances

"Smart" toys

Digital cameras and television

Transistor radios

Computer-controlled industrial machines

Robots

Air traffic control systems

ceramics play in the electronics revolution that has so dramatically changed our lives. Let's start by exploring what the term *electronics* means and by reviewing some of the historical milestones that have contributed to the electronics revolution. Then we'll discuss some of the magical ways that ceramics fit into this picture.

WHAT WE MEAN BY *ELECTRONICS*

Electronics involves the control of electricity to create all of the magic that goes on inside our computer, our CD player, the electronic-ignition system in our car, and myriad other pieces of electronics equipment. Remember the electrons we talked about in Chapter 4—the tiny negatively charged particles that orbit around the nucleus of each atom? These are the primary carriers of electricity. We harness these electrons for our use by forcing them to flow through metal wires, just like we force water to flow through pipes and hoses. In fact, we can picture in our minds what happens inside electrical devices by a comparison with the ways we harness and control water, such as by flowing, storing, and pumping it.

Let's start our comparison at a reservoir. Just as a reservoir is a source of water, electrical generating plants or batteries are sources of electricity. The weight of the water in the reservoir pushes water into pipes or canals to be transported to our city. The pressure or force that pushes electricity is the *voltage* that's built up at the generator or stored in the battery. Another term electrical engineers use for voltage is *electromotive force* (emf). The emf pushes the electricity into metal transmission lines and provides the force to transport the electricity to its destination—our homes and electronics equipment. The amount of electricity that flows through the transmission lines is called the *current,* just like the rate of water flow in a river.

Depending on its use, water needs to be flowed at different rates. Sometimes, water flow must be restricted to only a tiny amount, as in our water faucets. Other times, the flow rate must be increased by pumping, as for irrigation or many industrial processes. The same is true with electricity. A silicon chip—the brain in a computer—requires a different pressure and flow (voltage and current) of electricity than does a light bulb or an electric motor. The flow and pressure of water is controlled by the size of the pipes or channels through which it flows, and by devices such as pumps, gates, and valves. The pressure and flow of electricity is controlled by the size and degree of electrical conduction of the wire that carries it, and by devices such as *transformers, insulators, resistors, capacitors, semiconductors, transistors,* and *diodes.* Nearly all of these electrical devices require ceramics. As we continue through this chapter, we'll come back to some of these devices to explore the role of ceramics in them and the importance of the devices to the electronics revolution.

Ceramics can be used in many different electrical devices because there are so many different kinds of ceramics: at least 3000. Some of these ceramics completely block electricity and are called *insulators,* like the alumina spark plug insulator that we discussed in Chapter 2. These ceramics can be compared to the pipe and the hose that contain water. Just as water can't leak through these structures, electricity can't leak through a ceramic insulator.

Other ceramics, called *resistors,* allow only some electricity to pass through, similarly to pinching a hose or straw to decrease the amount of water that gets through it. The resistor controls the amount of electricity that gets through a wire. Just as you have to suck harder to get water through the pinched straw, the electricity has to work harder to get through the resistor. This extra work produces heat, and in some cases a lot of heat. Resistive materials make up the burners on your electric stove, the heater in your toaster, and the very high-temperature ceramic heaters that melt glass for producing windows and bottles, and silicon for making computer chips.

CERAMICS RESPOND IN DIFFERENT WAYS TO ELECTRICITY

Electrical insulator	Prevents electricity from passing through a wire
Capacitor	Stores electricity and filters out some electrical signals
Resistor	Allows electricity to get through, but with difficulty
Semiconductor	Acts as a special "gate" that controls the flow of eletricity
Electrical conductor	Allows electricity to pass through
Superconductor	Allows electricity to pass through with no resistance
Piezoelectric	Converts between pressure and electricity
Silicon chip	Supplies the "brains" of a microprocessor for a computer integrated circuit or other electronics devices
Thermistor	Exhibits change in electrical resistance with change in temperature
Varistor	Exhibits change in electrical resistance with change in voltage

An electrical *conductor* is a material that allows electricity to pass through it easily, with only a tiny amount of resistance. Most metals are good electrical conductors. Remember our discussion in Chapter 4 about the electrons that bond metal atoms together? These electrons aren't attached to a single atom or even to a pair of atoms next to each other. Instead, they're free to move from atom to atom throughout the piece of metal. These are the electrons that carry electricity through a metal wire. In contrast, the electrons that bond most ceramic materials together aren't free to move throughout the piece of ceramic, because they're tied up between pairs of neighboring atoms. These ceramics are electrical insulators.

We also talked a little in Chapter 4 about *semiconductors*. As you may remember, semiconductors are similar to ceramic insulators but have a small energy gap, or band gap. If enough energy is applied to such a material, electrons can jump over this energy gap, and the material becomes an electrical conductor. Until this extra energy is added, however, the material acts like an insulator. The semiconductor is like a gate that controls the flow of water. The gate in an irrigation ditch can be moved up or down, at the choice of the gatekeeper, to either block the water or let it flow at the rate the gatekeeper selects. A semiconductor device can do the same thing with electricity, as we'll discuss later, when we talk about the silicon chip and integrated circuits for our desktop computer.

We'll also talk later about other important electrical characteristics of ceramics, such as piezoelectricity and superconductivity; but now let's review some of the historical milestones in our understanding of electricity and in the dawn of the Electronics Age.

MILESTONES IN OUR UNDERSTANDING AND USE OF ELECTRICITY

Materials and Electricity

Some of the effects of electricity have been known for many centuries. The earliest humans witnessed the power of lightning but interpreted it as an act of the gods. Classical Greek scholars observed that amber (fossilized tree resin) rubbed with a piece of cloth attracted lint and little pieces of paper and also made a person's hair stand on end. They believed there was a mystical force present, but they couldn't figure out what it was. Today, we know that amber is a natural plastic that builds up and stores an electrical charge when rubbed.

No one really started to understand electricity until the late 1700s. We've all heard of Benjamin Franklin and his kite. This was the time frame for the first controlled scientific experiments to trap, harness, generate, and manipulate electricity. These experiments continued through the 1800s and resulted in important inventions such as the electricity generator, the electric motor, electricity transmission through wires, and the electric light. Amazingly, all were invented without the inventors knowing or understanding the source or cause of electricity! The electron and the basic principles of electricity weren't discovered until 1897, by Joseph John Thompson. In fact, Thompson didn't even come up with the name for electrons; he wanted to name this new subatomic particle a "corpuscle." Instead, though, the scientific community selected *electron,* the ancient Greek name for amber.

Even though scientists didn't know the exact source of electricity, they learned that they needed materials that could conduct this mysterious force and block its flow or keep it from leaking. Metals were identified as good conductors, and they also could be formed into long wires because of their special properties of ductility and malleability. As discussed in Chapter 2, ceramics and glass were identified during the earliest experiments as excellent insulators for electricity and were used throughout the 1800s to support electrical invention and innovation.

Scientists learned other things about electricity in the 1800s. They learned that the force could work either by flowing in one direction around a circle (a circuit), which became known as *direct current* (dc), or by rapidly alternating the direction of flow, which became known as *alternating current* (ac). The batteries in our cars and in our flashlights produce direct current. Electrical generating plants (nuclear, hydroelectric, and coal-burning) produce alternating current, which has been standardized in the United States to alternate direction 60 times each second. Alternating current produced at the generator has very high voltage (pressure) but low current, like water in the lines that enter your house. The water is just sitting there under pressure, until you turn on the faucet. Similarly, the alternating current is just sitting there, instantly available for you as you turn on a light or your electric can opener.

Some electrical devices require dc and others ac. Scientists discovered, in 1870, that the naturally occurring ceramic *galena* (a chemical compound of lead and sulfur) was a semiconductor and could convert ac to dc. Essentially, the galena acted like a one-way street or tunnel. The ac electricity could enter and pass through the galena during the first half of the alternating cycle, but it couldn't reverse and come back when the ac cycle was reversed. The electricity entering the galena was ac and that exiting the galena was dc. This discovery allowed the scientists to invent new electrical devices. One of the most important devices allowed radio waves to be converted into electrical pulses, which could then be converted into sound: the first step in the development of the radio.

The first significant radio transmission and reception was by Guglielmo Marconi, in 1896. He sent a wireless message a distance of 1.5 miles (2.5 kilometers). By 1901, Marconi had succeeded in sending radio communication across the English Channel and then across the Atlantic Ocean, but the technology was not yet practical for use by the general public.

ELECTRONICS MILESTONES

1897 Discovery of the electron

1902 Invention of the vacuum tube

1947 Invention of the transistor

1958 Invention of the integrated circuit

After the discovery of the electron, the field of electronics really began to blossom. The first key development, between about 1902 and 1906, was the vacuum tube. The vacuum tube could do something new; it could amplify an electrical input or signal. Imagine how important this was to the development of the radio. One challenge for radio technology was the need to receive a faint radio-wave signal and convert it into a strong enough electrical signal able to produce sound that could be heard easily by the listener. The vacuum tube filled this need.

Figure 6-1. *Vacuum tubes, showing the metal plates and wires encased in a glass "envelope." Largest tube shown is 4.5 inches high.* (Photo by D. Richerson.)

Most vacuum tubes consisted of an array of metal wires and plates sealed inside a glass tube, with metal prongs coming out the bottom for plugging into the electrical equipment. Before the tube was sealed, all of the air was sucked out, to form a vacuum inside the tube around the metal parts. A typical vacuum tube was around 1 inch in diameter and 2 to 3 inches long, resembling a miniature space ship.

By 1910, vacuum tube technology had made the radio practical but still not for the general public. These early radios were used primarily for communication between oceangoing ships and to the shore. Such communication became especially important during the First World War. After the war, though, the radio went public, marking the first widespread use of electronics. About $60 million worth of radios were sold in 1922. By 1929, radio sales had mushroomed to $900 million. The Electronics Age was on its way! Radio was followed, in the late 1920s, by the first black and white television and, in the 1940s, by the first color television. It's amazing to realize that radio and television have only been around since the time of our grandparents or parents.

The vacuum tube was a true milestone in electronics, but it had its limitations. Vacuum tubes were bulky and fragile, required a lot of electricity, and produced so much heat that they required cooling. Because of the size of the vacuum tubes and other electrical devices needed for a radio, early radios were quite large—at least as large as the monitor for your desktop computer. All other early electronics equipment was also quite large.

The Invention of the Transistor

Scientists continued to improve their understanding of electrons and electricity. In the late 1920s, Felix Bloch proposed the *band theory*, discussed briefly in Chapter 4. The theory suggested that the electrons in materials are in energy bands. In some materials, especially metals, key bands overlap, and electrons can flow through the material to provide electrical-conduction behavior. In other materials, including most ceramics and organics, the bands are widely separated (a wide band gap), and the material is an electrical insulator (highly resistant to the flow of electrons). And in some ceramic materials the key bands (the valence band and conduction band) are close enough together (a small band gap) that a small external energy source, such as heat or light, can boost electrons from the valence band to the conduction band. These materials are semiconductors.

Figure 6-2. Comparison of the size of a vacuum tube with two types of transistors, one encased in metal and the other in plastic. (Photo by D. Richerson.)

The band theory convinced scientists that a solid-state (small, simple, one-piece) device constructed from semiconductor materials could do the same thing as the vacuum tube. This discovery started a quest that ultimately would change electronics and our lives forever. The first success in the quest was achieved at Bell Laboratories in 1947/1948, by John Bardeen and Walter Brattain. They placed a solid piece of semiconductor made of *germanium*, a silicon-like chemical element, in a special way between metal contacts and demonstrated that the simple device could amplify electricity in the same way a vacuum tube did. They referred to their device as a *transfer resistor,* which later became commonly known as a *point contact transistor* or simply a *transistor.* Within a couple of years, their colleague William Shockley invented an improved transistor. The three scientists received the Nobel Prize for physics in 1956.

The transistor was a major breakthrough, allowing electronic devices and equipment to be much smaller than was possible with vacuum-tube technology. Besides being much smaller, the transistors used a tiny fraction of the electricity required for a vacuum tube and gave off much less heat. The transistor radio (pocket radio) is a good example of the impact the transistor and its capability for miniaturization had on the general public. For the first time, a radio could be made small enough, and with a low enough power requirement, to be battery operated and totally portable.

Since 1950, scientists have invented new types of transistors that can do many things besides amplifying electricity. Some can act as tiny switches and gates to control the amount of electricity flowing through each electrical device in the electrical circuit inside electronics equipment. Some can convert from ac to dc. Others can be arranged to store information inside a computer.

The Invention of the Integrated Circuit

The transistors we've discussed so far were great inventions but still didn't make possible hand-held calculators, desktop computers, and all of the electronics we use today. First, the transistors and other elements of an electrical circuit (conductors, resistors, insulators, and so forth) had to be reduced in size by thousands of times, until they were small enough to fit onto the head of a pin. The technology that made this reduction possible was the *integrated circuit* (IC). The first IC was demonstrated by Jack Kilby, of Texas Instruments, in 1958. This early IC was really simple compared to ICs available today. It consisted of a single transistor combined with a resistor and a capacitor (we'll discuss capacitors later), all built into a tiny slice of semiconductor silicon. We now commonly refer to such an IC as a *silicon chip* or simply a *chip.* About the same time, Robert Noyce and Gordon Moore, of Fairchild Semiconductor Company, invented a similar IC and came up with ideas for how to build it into useful electronics devices.

By 1962, the IC technology had been scaled up to chips about one-tenth of an inch square, containing about ten electrical components (transistors, resistors, and so forth). About $4 million worth of these simple ICs were sold in 1962, hardly an auspicious start. However, the technology

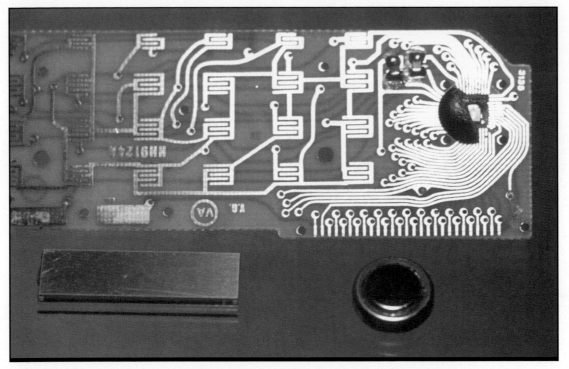

Figure 6-3. *Silicon chip "brain" in a small hand-held calculator. The chip was mounted onto a glass-fiber-reinforced plastic strip, connected by tiny wires to conductive metal circuits, and potted (enclosed) in a protective layer of plastic. Part of the plastic has been cut away to reveal a portion of the chip. The battery and liquid crystal display, both containing ceramic materials, also are shown.* (Photo by D. Richerson.)

advanced rapidly, especially the ability to put more electrical components on each silicon chip. By the early 1970s, silicon chips one-tenth of an inch square each contained thousands of electrical components and were complex enough to be the brain inside a hand-held calculator. This electronic innovation became known as large-scale integration (LSI). The market reached around $2 billion by 1974.

From then on, technology progressed at breathtaking speed. Very large-scale integration (VLSI), with more than 3 million transistors on a silicon chip, was introduced in the 1980s. Such a chip, no larger than your thumbnail, could be designed with all of the electrical circuits necessary for a computer that previously had filled a whole room. This chip, known as a *microprocessor,* is the magic inside (the "Intel Inside") our desktop computers and all of the other electronics equipment we use every day, including our fax machines, photocopiers, printers, microwave ovens, automobile electronics control systems, digital cameras, and heart pacemakers. Now chips with 16 million transistors are commonplace. Ceramics are absolutely essential in these chips and to the many steps involved in making them, as we'll discuss in the next section.

THE ROLE OF CERAMICS IN IC TECHNOLOGY

Inside a Silicon Chip

How do scientists fit 16 million transistors, plus all of the electrical conducting paths, insulators, resistors, and other electrical devices, into an integrated circuit the size of a fingernail? They start by making each device incredibly tiny, about one twenty-millionth of an inch across (about 500 nanometers). These devices are built up on the surface of a slice of single-crystal silicon, layer after layer. A single chip can have more than 20 layers, all interconnected like a three-dimensional maze.

To imagine the complexity of an IC, think of a 20 story building filled with thousands of apartments. Each apartment is full of electrical wiring and different appliances such as refrigerators, toasters, and hair dryers. These appliances are like the individual devices in the microchip, each requiring its own special electrical connections and input, yet all interconnected to the main electrical wiring of the whole building. Now, imagine the network of pipes and plumbing for distributing water; this is like another network of devices, forming a separate circuit but also necessary to the overall duties of the integrated circuit. Finally, imagine all of the hallways and doors. These represent another circuit.

Now shrink all of the electrical wiring, plumbing, halls, electrical switches, faucets, and doors in our 20 story building down to the size of a fingernail, but squashed and compacted to leave no open spaces. This incredibly compact structure is a pretty good illustration of the complexity of a modern IC. The electrical wiring, plumbing, and halls represent the complex electrical paths in the IC. The electrical switches, faucets, and doors represent all of the microswitches and control gates that route the electrical current through the IC. The millions of appliances represent the individual electrical devices built into each IC. Engineers who design ICs select which devices to build in, according to the desired use of the IC: in a desktop computer, a calculator, a talking toy, or a breadmaker. We'll discuss soon how engineers make an IC, but let's talk a little about the materials inside an IC, especially the ceramics.

Most of the materials in an IC are semiconductors, meaning they can block electricity under some circumstances and conduct electricity under other circumstances, as discussed earlier. They can act as tiny microswitches to route electricity through the IC to perform tasks such as storing information (computer memory) or going through a complex sequence of mathematical calculations (calculator or computer).

Some materials, such as silicon and germanium, are natural (*intrinsic*) semiconductors, because they have a narrow band gap; not much energy is needed for electrons to jump the energy gap and let electricity flow. Some materials, including silicon and germanium, can be made into special "designer" (*extrinsic*) semiconductors by adding tiny amounts of other chemical elements, such as boron and phosphorus. This addition is called *doping*. Doping with boron makes extra electrons available in the silicon, so that electrical conduction can occur more easily by the flow of these extra negative charges. Doping with phosphorus causes a deficiency in the number of electrons, which is the same as making positive charges (*electron holes*) available for conduction. By using natural and doped semiconductors, engineers can design just about any electrical behavior they want. However, they must also separate the different electrical circuits and devices from each other with an electrical insulator. This separation is an important role for thin films, or layers, of ceramics.

We identified three basic categories of materials at the start of this book: metals, ceramics, and organics. Into which of these categories do semiconductors used in ICs fit? They're clearly not organic, but the choice between metal and ceramic is more difficult. Silicon, for example, is shiny and looks metallic, but the atoms are bonded together in exactly the same way as in diamond, which is usually used as one of the classical examples of a ceramic structure. Germanium also has the same crystal structure as diamond. So, from an atomic-structure point of view, these semiconductor materials are more like ceramics. If we can reasonably consider these materials as ceramics, then most of an IC chip is made up of ceramics.

Making Integrated Circuits with Ceramics

The fabrication of ICs is a $150 billion-per-year business and the heart of the huge $15 trillion electronics market. Ceramics are involved in just about every step in making ICs. The following paragraphs describe briefly how an IC is fabricated and the role of ceramics.

The first step in IC fabrication is to produce a very pure, perfect single crystal of silicon. Remember our discussion in Chapter 2 on the growth of ruby crystals by the Verneuil method, by dropping alumina powder through a high-temperature torch and building up a

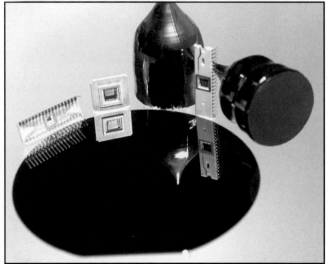

Figure 6-4. *Two single-crystal silicon rods, showing 1960s fabrication capabilities; a large polished wafer about 8 inches in diameter, showing early 1990s capability; and several chips mounted in ceramic packages. Note the mirror polish on the large wafer. (Large wafer provided by MEMC Southwest, Sherman, TX. Photo by D. Richerson.)*

single crystal boule? This method was tried with silicon but didn't result in a perfect enough crystal. Gordon Teal and John Little of Bell Laboratories developed a different method that worked. They melted the silicon in a container (crucible) made of high-purity fused silica ceramic (the material that Hale initially tried for making telescope mirrors), which was very stable and didn't contaminate the silicon at the melting temperature, above 2560°F. Once the silicon was melted, a single crystal of silicon was touched to the top surface of the melt, allowing silicon atoms from the melt to deposit. If the silicon crystal was very slowly pulled away from the molten silicon, a single crystal *rod* of silicon was produced, similar to the one shown in Figure 6-4.

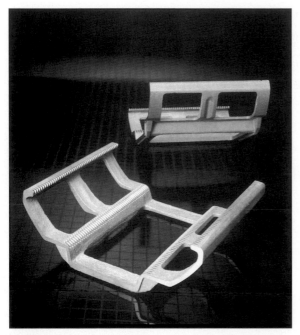

Figure 6-5. *Silicon carbide wafer boats (above) and a wafer boat loaded with silicon wafers and supported on a silicon carbide paddle (below).* (Photos courtesy of Saint-Gobain/Norton Industrial Ceramics, Worcester, MA.)

In the early 1960s, when ICs were first being developed, the largest single crystal of silicon that could be grown was about 1 inch in diameter. The technology has come a long way. Now, highly perfect single crystals of silicon can be grown 12 inches in diameter and 40 inches long.

The second step in IC fabrication is to slice the single-crystal rod into wafers about 25 thousandths of an inch thick and to grind and polish these wafers to an incredible degree of flatness and smoothness (note the mirror finish of the wafer in Figure 6-4). This shaping is necessary because the devices that will be formed later at and on the surface of the wafer are only 20 millionths of an inch (500 nanometers) across. Even the tiniest scratch or pore will disrupt the formation of

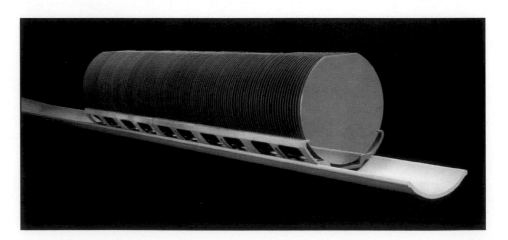

these devices. All the precision slicing, grinding, and polishing are done with ceramics.

The polished silicon wafers are carefully cleaned to remove even the tiniest particle of dust or lint, loaded onto a ceramic carrier called a *wafer boat,* placed on a long-handled flat ceramic paddle, and inserted into a ceramic-lined furnace. The boat, paddle, and furnace lining usually are made of silicon carbide, because it can withstand the temperature in the furnace and doesn't contaminate the silicon. The desired dopant to give the silicon wafer just the right semiconductor characteristic is also loaded into the furnace. When the furnace is heated to the right temperature, the dopant diffuses (soaks) into the surface of the silicon.

We now have the silicon base or *substrate* for our chip, but we still have to build the many layers of transistors and other devices and the ceramic electrical insulators separating them. This layering is done by repeatedly etching (dissolving) away a microscopic pattern of some of the silicon and depositing other ceramic materials in its place, layer by layer.

The etching is done using a special process known as *photolithography.* The silicon is coated with a light-sensitive polymer. Light is projected through a lens and a mask (a cutout pattern) to shine the pattern of the particular layer of devices onto the silicon chip. The polymer struck by the light becomes sensitized, similarly to the Fotoform glass discussed in Chapter 4. The sensitized polymer and a tiny depth of silicon underneath can now be etched away by an acid. Since the devices must be accurate to about 20 millionths of an inch, the mask and lens must be very accurate. The lens, for example, is made of low-thermal-expansion fused silica glass and costs about $2 million.

Figure 6-6. *Vacuum chuck used to hold a silicon wafer in position during various high-temperature vapor deposition steps. Approximately 12 inches across, showing outline of position of silicon wafer.* (Photo courtesy of Saint-Gobain/Carborundum Structural Ceramics, Niagara Falls, NY.)

The etching step is followed by a deposition step, to fill the device material into the spaces etched out of the silicon. Sometimes, this deposition is done by chemically reacting gases to deposit the desired material; other times, it's done by *sputtering,* knocking atoms off the surface of a target and letting them redeposit on the etched silicon wafer. These are all high-temperature processes that require ceramic furnace linings and support structures.

The photolithography, etching, and deposition processes are repeated with different materials for each type of device in each layer, until all of the layers of the IC have been formed. This process is amazing and complex.

Figure 6-7. *Close-up of a 1960s silicon wafer after deposition of devices, with the tip of a sewing pin for size comparison.* (Photo by D. Richerson.)

Any mistake made in any step ruins the whole wafer. Since each wafer contains hundreds of individual ICs, a ruined wafer is an expensive loss.

A complete description of the design and fabrication of ICs would take many pages and go into scientific details well beyond the scope of this book. The intention of this brief description was to show you how amazing the silicon chip really is and also to show that the silicon chip—and all of the electronics products made with silicon chips—wouldn't be possible without ceramics.

Now What Happens with the Silicon Chip?

Now that we have a silicon chip, what do we do with it? How is it mounted into a piece of electronics equipment? We already saw one example in Figure 6-3. The silicon chip was mounted onto a thin sheet of plastic reinforced with glass fibers. The chip next was connected by tiny wires to a pattern of metal circuits painted onto the plastic and leading to the battery, keyboard, and liquid-crystal display. A liquid polymer then was poured over the chip and wires and allowed to harden, just like an epoxy glue starts out as a liquid and then hardens: A silicon chip can be damaged by moisture in the air and so must be completely encapsulated.

Plastic encapsulation works fine for small ICs in calculators, but not for larger ICs in computers or other equipment that runs continuously. The ICs build up a lot of heat, enough to actually ruin the IC. Since plastic holds in heat, an alternate material must be used, one that can carry the heat away from the chip. Aluminum oxide is the material most commonly used, but not only to protect the chip from moisture and carry away heat. Engineers have found a way to build metal wiring right into the alumina to form a solid *hybrid package,* a wonder of miniaturization that rivals the miniaturization in the IC. A cross-section diagram of a hybrid package is shown in Figure 6-8. The hybrid package consists of many layers of alumina, which is an excellent electrical insulator, interlaced with all of the metal electrical-conduction paths necessary to carry electricity to all of the complex circuits in the IC. Let's explore how this hybrid package is made and used in a mainframe computer.

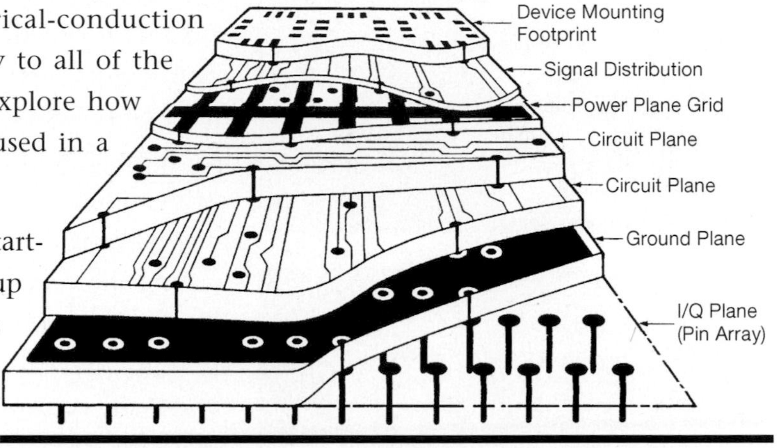

The hybrid package is made starting with alumina powder ground up as fine as dust. The powder is mixed with a liquid and a selected organic *binder* material (sort of like white household glue) that will become solid but flexible when dried. This mixture, called a *slurry* or *slip,* is a fluid like buttermilk or eggnog. The slip is poured onto a large, flat surface and spread out with a knife blade to a flat, thin layer and allowed to dry. After it's dried, the material is a mixture of ceramic particles and the flexible organic material

Figure 6-8. *Schematic illustrating the construction of a hybrid multilayer ceramic package for mounting silicon chip integrated circuits* (from E. Ryshkewitch and D.W. Richerson, Oxide Ceramics, 2nd Edition, General Ceramics, Inc., Haskell, NY, 1985.)

Device Mounting Footprint
Signal Distribution
Power Plane Grid
Circuit Plane
Circuit Plane
Ground Plane
I/Q Plane (Pin Array)

and referred to as a *tape.* It looks a little like the very thin dough rolled out for making noodles or fine French pastry but is even thinner and very smooth on both surfaces.

The next step is to cut the tape to the desired size and punch tiny holes where metal conductive paths through the layer are required. A paste of metal particles (typically tungsten or molybdenum) mixed with an organic binder and a little liquid is printed onto the tape in the type of circuit patterns shown for the different layers in Figure 6-8. This metal paste also fills the holes that were punched. The different layers of tape, with their printed metal patterns, are now stacked and gently heated under slight pressure, to cause

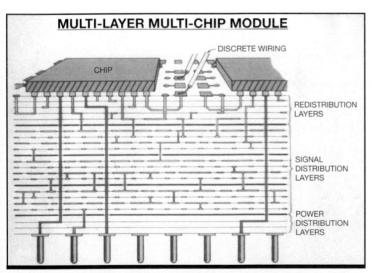

MULTI-LAYER MULTI-CHIP MODULE

DISCRETE WIRING
CHIP
REDISTRIBUTION LAYERS
SIGNAL DISTRIBUTION LAYERS
POWER DISTRIBUTION LAYERS

Figure 6-9. *Schematic showing the complexity of a 23 layer hybrid multilayer ceramic package.* (Courtesy of IBM, Hopewell Junction, NY.)

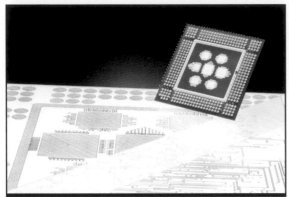

Figure 6-10. *Examples of hybrid multilayer ceramic packages. (Photos courtesy of Kyocera.)*

the organic binder in the tape to bond the layers of tape together. The large ICs for mainframe computers require more than 20 tape layers.

The layers of tape are still just a bunch of ceramic and metal powder particles held together with an organic glue, just as if you mixed sand or dust with white glue. Before the hybrid package can be used for ICs, it must be converted to a solid ceramic interlaced with conductive metal channels. This is done by carefully burning off the organic binder material and then heating the ceramic to a very high temperature (typically over 2900°F). If heated correctly, the resulting ceramic contains no pores, and the internal metal "wires" are continuous. As shown in Figure 6-8, there's even a recessed area built into the package, sized just right to hold the IC.

The hybrid IC package is a miracle of miniaturization that required many innovations in materials and fabrication technology. A hybrid package built in the 1960s could hold only one chip and provide the internal electrical wiring for only a few hundred devices.

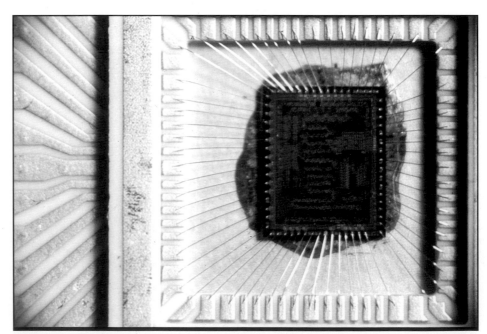

Figure 6-11. *Magnified view of a vintage 1975 silicon chip brazed into a ceramic package. Note the faint image of the devices on the silicon chip and the tiny wires connecting the chip to the ceramic package. (Photo by D. Richerson.)*

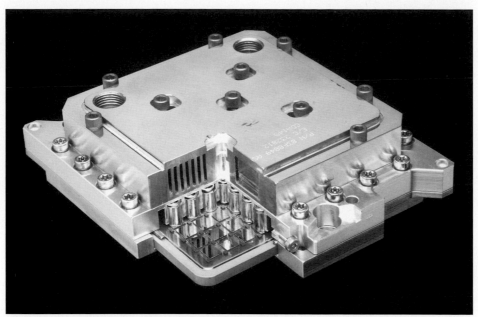

Figure 6-12. *Cutaway of an IBM mainframe computer thermal conduction module, showing silicon chips on a hybrid IC package. Helium is circulated through the metal housing to carry away heat produced by the ICs.* (Photo courtesy of IBM Microelectronics, East Fishkill, NY.)

Now, hybrid packages such as the IBM mainframe computer module illustrated in Figures 6-12 and 6-13, can house 133 powerful chips, each containing millions of devices. A 1988 version of this IBM module was about three and one-half inches square and one-fifth of an inch thick. It contained 33 layers, about 350,000 metal-filled holes between layers, 427 feet of internal metal wiring, and 1800 brazed-on pins for external connections. The IBM module contained over 100,000 separate electrical circuits in all and generated so much heat that it required an elaborate cooling system. This hybrid package, plus the 133 silicon chips, replaced a whole room full of equipment required to do just a fraction of the same computations 20 years earlier.

OTHER ELECTRICAL USES FOR CERAMICS

As you can now see, integrated circuits and hybrid packages wouldn't be possible without ceramics. These technologies have profoundly affected our lives, in everything from our personal computers to the mainframe computers that handle transactions of banks, airlines, and the Internet. As important as ICs and hybrid

Figure 6-13. *IBM mainframe computer IC package. Bottom and top view of IC package, showing the 1800 pins for electrical connections and the 121 logic and array silicon chips. The package and chips fit into the IBM thermal conduction module.* (Photos courtesy of IBM Microelectronics, East Fishkill, NY.)

packages are, though, ceramics have many other important electrical applications. Let's explore some of these other applications.

Ceramic Electrical Insulators

Ceramics and glass have been used as electrical insulators since the discovery of electricity. Glass and porcelain were commonly used during the 1800s, and—as you learned in Chapter 2—alumina became important by the mid 1900s for spark plugs in motor vehicles and airplanes. Since then, these and other ceramics have been used as electrical insulators in just about every device that operates on electricity. You'll find them in radios and televisions, all household appliances, and nearly all motors, communications equipment, microwave systems, computers, industrial and medical equipment, and aerospace and military systems.

One of the oldest uses of ceramic insulators is to insulate power lines that transport electricity from the hydroelectric dam or power-generation plant to our homes. Next time you drive through the countryside, take a closer look at the high towers that support the high-voltage electrical-transmission wires. The wires never touch the towers but instead are attached to large ceramic insulators. These ceramic insulators must be able to withstand very high voltage and also must be strong enough to hold the weight of the wires, even during a heavy wind or snow storm. Once the electricity reaches our home, ceramic insulators line electrical outlets

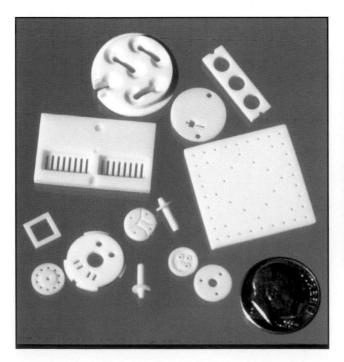

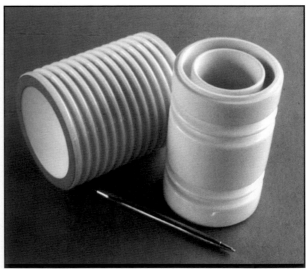

Figure 6-15. *Medium sized electrical insulators of aluminum oxide.* (Samples from Coors Ceramics, Golden, CO. Photo by D. Richerson.)

Figure 6-14.(left) *Small aluminum oxide electrical insulators for various electrical applications fabricated by Western Gold and Platinum, Belmont, CA, circa 1980.* (Photo by D. Richerson.)

and light sockets, not only to protect us from getting shocked but also to give extra protection against fire.

Another important use of ceramic insulators is for substrates or mounting brackets to which other electrical components are attached. The plastic reinforced with glass fibers shown earlier, in Figure 6-3, is an example of a substrate, and many of the small parts shown in Figure 6-14 are ceramic mounting brackets. As mentioned before, ceramics are often used because they can carry heat away more rapidly than plastic can. Alumina works fairly well and is used, for example, to mount all of the electronics systems in our automobiles. Some products, such as some laser parts, require ceramic materials that carry away heat even better than alumina can. Beryllium oxide and aluminum nitride both conduct heat away at least four times faster than alumina—nearly as fast as metals. But the champion is diamond: Diamond carries away heat five to ten times better than do the best metals.

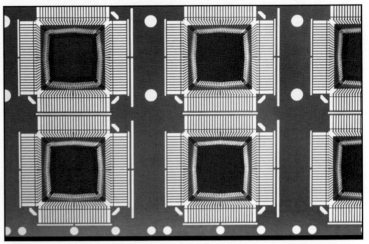

Figure 6-16. *Diamond substrates in copper frames. Diamond and copper are superior heat conductors and effectively carry heat away from high-power silicon chip integrated circuits.* (Sample from Saint-Gobain /Norton Advanced Ceramics, Northboro, MA. Photo by D. Richerson.)

Dielectric Ceramics and the Capacitor

Ceramic *capacitors* are electrical devices required in virtually every piece of electrical equipment. More than one billion are produced each day. The capacitor is actually a special type of electrical insulator called a *dielectric*. Even though it won't let electrons flow through, a dielectric is still affected by electricity. One side of the ceramic becomes more positive and the other side more negative, which is where the name dielectric ("two-electric") comes from. This movement of electrical charge to opposite sides (poles) of the ceramic is called *polarization,* and is illustrated in Figure 6-17. It allows the ceramic to store electricity, which is one of the important tasks of a capacitor.

A capacitor is made up of alternating layers of ceramic and metal. When electricity is passed through the metal layers, the ceramic becomes polarized. Some ceramics work better than others. Alumina shows very little polarization, which is why it's such a good insulator. A special ceramic called *barium titanate* (made up of atoms of barium, titanium, and oxygen) polarizes thousands of times more than alumina and is very effective in

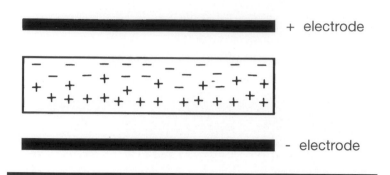

+ electrode

- electrode

Figure 6-17. *Schematic illustrating the shift in distribution of electrical charge in a dielectric ceramic during polarization. Note that the positive charges move toward the negative electrode and the negative charges move toward the positive electrode.* (Illustration from D.W. Richerson, Modern Ceramic Engineering, Marcel Dekker, NY, 1992.)

capacitors. In fact, barium titanate has played an important role in miniaturization. The radio is a good example. A typical transistor pocket radio in the 1970s contained around 10 capacitors made with barium titanate. Each of these capacitors was around three-eighths of an inch in diameter. In 1940, the best dielectric ceramic for capacitors had only one five-hundredth the polarization of barium titanate. A capacitor made with this material would be much too large to fit into a portable radio.

Capacitors can do other tasks besides storing electricity. Some can block direct current and allow alternating current to pass; others allow some frequencies of electromagnetic signals—such as radio waves—to pass, and block others. Take a look at the dial of your radio. The numbers you see represent different frequencies of radio waves. Early radios used capacitors to allow the listener to "tune in" to the various broadcast stations. New radios use even better dielectric ceramics.

Cellular Phones

Another important use for dielectric ceramics is in cellular phones, where they act as a *block filter*. A cellular phone receiver is constantly bombarded by unwanted signals (radio waves, TV waves, and so forth), as well as the telephone signals we want to receive. Each of the signals, wanted or not, has a different wavelength and frequency (just as colors of light all have different wavelengths and frequencies). The

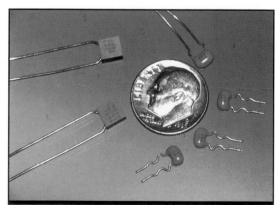

Figure 6-18. *Multilayer ceramic capacitors. Multilayer ceramic capacitors are fabricated similarly to multilayer packages. Layers of dielectric ceramic tape are coated with a thin layer of conductive metal and stacked. The whole assembly is fired to produce a miniaturized capacitor.* (Samples from AVX Ceramics Corporation, Myrtle Beach, SC. Photo by D. Richerson.)

dielectric ceramic in a cellular phone has been designed to interact with these frequencies to allow the telephone message to pass, while "sifting out" all of the other frequencies. This wasn't an easy task to accomplish. The first successful block filter was over 8 inches long, definitely too big

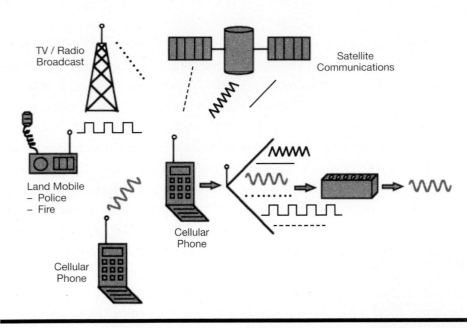

Figure 6-19. *The role of the ceramic block filter in a cellular-phone. The cellular phone antenna picks up multiple signals from many wireless communication devices. The ceramic block filter selects the desired signal by preventing undesired signals from entering the receiver of the phone. (Illustration prepared by John Estes, Motorola Component Products Group, Albuquerque, NM.)*

and heavy to be used in a mobile phone. Block filters only became small enough to use in cell phones in the 1980s, but cellular phones are everywhere now.

Ceramic Superconductors

You've probably read in the newspaper about ceramic *superconductors* sometime since the first one was discovered, in 1987. The scientific community was excited

Figure 6-20. *Comparison of an early block filter (top, made mostly of metal) with later miniaturized ceramic block filters. (Photo courtesy of Motorola Component Products Group, Albuquerque, NM.)*

because the new ceramic superconductors had the potential to operate at much lower cost than previous, metal-based superconductor materials. The new ceramics offered new hope for a large reduction in electrical loss, or wastage, during transmission and storage of electricity; for low-cost nuclear magnetic resonance (NMR) medical scanners affordable to more people for the diagnosis of cancer and other medical problems; for high-speed trains that could ride on air, levitated by magnets; for highly efficient motors; and for incredibly fast supercomputers.

What's so special about a superconductor? A superconductor has absolutely no resistance to the flow of an electrical current. As we discussed earlier, most materials—even the best metallic conductors, copper and silver—have some resistance to the flow of electricity. The electricity has to work a little to get through these metals, which wastes some of the electrical energy and also produces heat. Electrical resistance is like friction. Tie a string to a small block of wood. Pull the block of wood across your cement driveway or a piece of sandpaper. It doesn't pull very easily, because there's a lot of friction. Now pull the block of wood across a sheet of very smooth glass. It pulls much more easily, because the friction is greatly reduced. This is like the flow of electricity through copper or silver. The resistance is greatly reduced, but still there. Now if you could bring the block of wood up on the Space Shuttle, take it for a space walk where there's no air or gravity, and push it off into open space, there would be no friction. This would be comparable to the flow of electricity through a superconductor, with absolutely no resistance or loss or heating.

Ceramic superconductors offer lower cost than earlier superconductors. Materials become superconductors only at a very cold temperature. The first superconductor, mercury, discovered in 1911, had to be cooled to about −460°F before it became superconductive. The only way to reach this incredibly cold temperature was to dunk the mercury in liquid helium, which is very difficult and expensive to produce and use. The ceramic superconductor discovered in 1987 caused so much

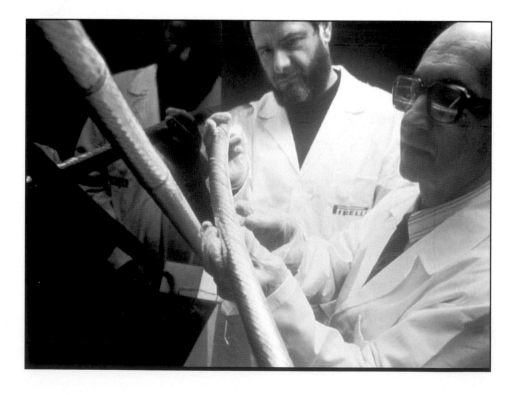

Figure 6-21. *Ceramic superconductor wire and experimental electrical transmission line.*
(Photographs courtesy of American Superconductor Corporation, Westborough, MA, and Pirelli Cavi S.p.A.)

excitement because it became superconductive at only –294°F, which is above the temperature of liquid nitrogen (–321°F). Liquid nitrogen can be prepared easily and sold at low cost.

A tremendous amount of research and development has been conducted on ceramic superconductors since 1987. One of the biggest challenges has been to make wire that's flexible and doesn't break easily. This has been achieved by mixing the ceramic with silver to form a composite wire that's ductile like a metal, yet retains the superconductivity of the ceramic when cooled with liquid nitrogen. As shown in Figure 6-21, experimental electrical-transmission line has been made with this ribbon-shaped wire. Superconductive ceramics also have been built into experimental motors, antennas, microwave devices, and magnetic shielding.

NATURE'S MAGNETS

Ceramic materials can also be magnetic. Magnetic ceramics are important in tape recorders and even in the recording tape itself, in loud speakers, in all of the motors that power the accessories in a car (windshield-wiper motors, power seats, power windows, and so forth), and even in the linear accelerators ("atom smashers") used by physicists to study the tiny particles that make up an atom.

Magnetism was first observed many centuries ago in the naturally occurring ceramic material *magnetite*. Ancient travelers learned to use magnetite as a compass for telling directions, but the force of magnetism wasn't understood until after the electron was discovered and scientists understood the structure of an atom. These scientists learned that electrons travel in pairs in orbits around the nucleus of the atom. The two electrons in each pair are spinning, similarly to the way the Earth rotates about its axis. This spin gives each electron a magnetic characteristic, a *magnetic moment*. However, the two electrons in an orbit spin in opposite directions, so that the magnetic moments are in opposite directions and cancel each other. If only one electron is present in an orbit, though, the magnetic moment doesn't cancel, and the atom shows magnetism. The primary chemical elements with these unpaired electrons are iron, cobalt, nickel, and manganese. As a result, some metals and ceramics containing these atoms are magnetic, like magnetite, which is made up of iron and oxygen atoms.

amazing facts

More than 1.3 billion pounds (nearly 600 million kilograms) of ceramic magnets were produced in 1997.

The modern age of ceramic magnets began in 1946, when J.L. Snoeck of Philips Laboratories in Holland synthesized ceramics with strong magnetic properties. He made these magnets similarly to the way that aluminum

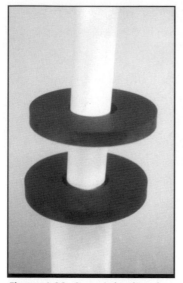

Figure 6-22. *Ceramic loudspeaker magnets. The magnetic field between the magnets is strong enough to suspend one in midair.* (Magnets from General Magnetic Company, Dallas, TX. Photo by D. Richerson.)

oxide and other polycrystalline ceramics are made: by starting with a powder, pressing the powder into the desired shape, and then firing the part to a high temperature. After they'd been fired, though, the materials still were not strong magnets. All of the little magnetic moments were pointing in every possible direction, mostly canceling each other. The secret found by Snoeck was to place the polycrystalline ceramic between two very strong magnets, which forced most of the tiny magnetic moments to line up in one direction, just like when you hold a magnet under a piece of paper with iron filings sprinkled on top: the iron filings all line up. The general term used for a ceramic magnet is *ferrite*.

Once some ceramic magnets have been magnetized, they retain their magnetism extremely well. These are referred to as permanent magnets or *hard ferrites*. Approximately 1.2 billion pounds (545 million kilograms) of hard ferrite ceramic magnets were produced in 1997 and used in motors for electric toothbrushes and knives, power accessories in automobiles (such as power seats and windshield-wiper motors), speakers, atom-smashing linear particle accelerators, and household magnets.

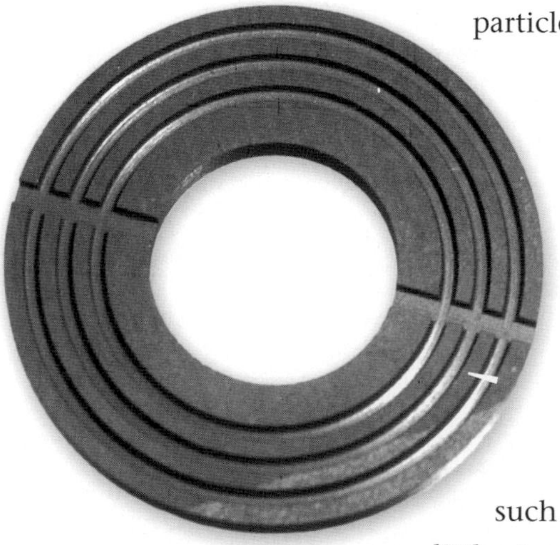

Other ceramic magnets can be magnetized and demagnetized readily by changing the direction of an applied magnetic field. These *soft ferrites* are important in television, radio, touch-tone telephones, communications, electronic-ignition systems, fluorescent lighting, high-frequency welding, transformer cores, and high-speed tape and disk recording heads. Approximately 120 million pounds (54.5 million kilograms) of soft ferrites were produced in 1997.

Figure 6-23. *Ceramic magnet for a microwave device.* (Photo by D. Richerson.)

Some of the earliest computers used tiny, soft-ferrite rings such as those shown in Figure 6-24 for storing information. Each little ring had three copper wires going through the hole in the middle. The flow of electricity through the wires was controlled so that each magnet could be independently programmed with the positive pole in either one direction or the opposite direction, which allowed storage of one piece of information. You can imagine how large and complex these early computers were.

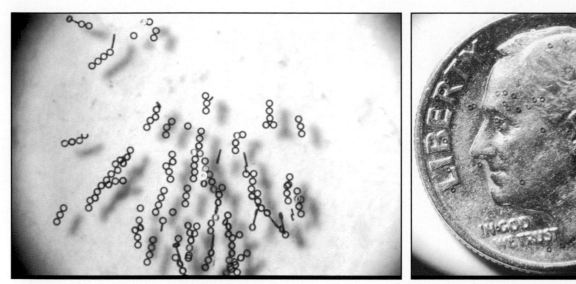

Figure 6-24. *Tiny ceramic magnets used in early computers, some shown on a dime for size comparison.* *(Photos by D. Richerson.)*

OVERVIEW: WHAT WILL THE FUTURE BRING?

You've probably been surprised by all of the electrical uses for ceramics and the dramatic effect they've had on our daily lives in this Age of Electronics. Ceramics will have an even bigger effect in the future. Computer capabilities are doubling every 18 months, as scientists and engineers learn how to make smaller and smaller integrated circuits and other electrical devices. We have amazing things to look forward to in the future, made possible by advanced electronics: powerful computers that will operate our industrial machines, communications systems, airplanes, and just about everything else better than they are today; more "smart" items, from toys to machines, that will interact directly with us through electronics; widespread use of virtual reality; cars that don't require drivers; medical devices built into our bodies or clothes, to continuously monitor our health, especially our heart; and maybe even computers that can communicate directly with our brain cells.

We left one of the most fascinating electrical characteristics of ceramics out of this chapter: piezoelectricity. Piezoelectricity defines the magical ability of some ceramics to convert back and forth between pressure—movement or vibration—and electricity. This unique property is playing an important role in allowing us to interact directly with machines and the world around us. The next chapter is dedicated to all of the incredible ways piezoelectric ceramics are already working for us.

Piezo Power

CHAPTER SEVEN

The wide range of ceramic electrical characteristics that we discussed in Chapter 6 is pretty amazing, but the most magical of all electrical properties of ceramics is *piezoelectricity. Piezo* means pressure, so piezoelectricity means "pressure electricity." Remarkable piezoelectric ceramics can actually convert back and forth between pressure and electricity. If you apply a pressure to a piezoelectric ceramic, even as slight as by breathing on it or talking to it, the ceramic will give off an electrical pulse. This behavior is used in microphones, underwater sound-detection hydrophones, and pressure- and vibration-detecting sensors. On the other hand, if you apply a voltage, the piezoelectric ceramic changes shape and either exerts a pressure on whatever it's touching or starts to vibrate. This reaction makes possible the quartz watch, the ultrasonic cleaner, medical ultrasound imaging, and an enormous number of other important products.

We'll discuss these applications and many others in this chapter. You'll probably be surprised to learn how often you encounter piezoelectric ceramics every day. Before we talk about applications, however, let's briefly review the history of piezoelectric ceramics.

PRODUCTS AND USES

Microphone
Loud speaker
Hydrophone sonar
Fish finder
Ocean floor mapping
Non-destructive inspection
Ultrasonic cleaners
Actuators
Motors, transformers
Sensors
Wheel balancers
"Smart" skis
Quartz watches
Buzzers, alarms
Musical instuments
Musical greeting cards
Lighters and igniters

THE HISTORY OF PIEZOELECTRIC CERAMICS

Some natural crystals are piezoelectric. Probably the most familiar to you are quartz and tourmaline. Quartz crystals are special to many who believe that these crystals can pick up good vibrations from the cosmos and channel them to benefit a person's health and well-being. Quartz also has been an important semiprecious gemstone throughout history, in the form of purple amethyst. Tourmaline also is an important gemstone, usually occurring in shades of red, pink, and green.

Figure 7-1. *Variety of piezoelectric parts and assemblies.* (Photo courtesy of Edo Corporation Piezoelectric Products, Salt Lake City, UT.)

The phenomenon of piezoelectricity was probably first observed by some Stone Age hunter making a stone axe or spear point, when he struck a piece of quartz with another stone and saw a spark. The first written description, however, was by Pierre and Jacques Curie, in 1880. They studied natural crystals of quartz and tourmaline and other crystals they grew in the laboratory from water-soluble salts. One was sugar and the other was called *Rochelle salt* and had especially strong piezoelectric behavior. You can easily grow small crystals of sugar at home. Simply stir a couple teaspoons of sugar into a couple tablespoons of water. The sugar grains will get smaller and smaller and disappear, but the sugar is still present, dissolved in the water. Set this sugar-water solution aside for a few days in a shallow, uncovered dish. The water will evaporate, and sugar crystals will form on the bottom of the dish.

Although the Curies learned about the magical ability of piezoelectric crystals to convert between pressure and electricity, they didn't come up with any practical uses for the property. Practical applications did start to show up in the early 1900s. For example, scientists learned that different-sized slices of quartz vibrated when exposed to different frequencies of radio waves, which led to an improved way to tune radios to different broadcast channels. A little later, piezoelectric crystals were used in phonographs.

By 1930, thin slices of Rochelle salt were used in microphones. The piezoelectric ceramic was attached to a diaphragm, like the top of a drum, but much smaller. When a person spoke into the microphone, the sound waves from his or her voice caused the diaphragm to vibrate at the same wavelength and frequency as the sound waves, much the way your ear drum vibrates to the sound waves hitting it. Low sounds have longer wavelengths (more distance between their peaks) than high sounds and thus fewer vibrations per second. You can demonstrate this for yourself by placing your fingers against your Adam's apple (larynx) and humming first a high pitch then a low pitch. You can feel that the vibrations are smaller for the high-pitched hum.

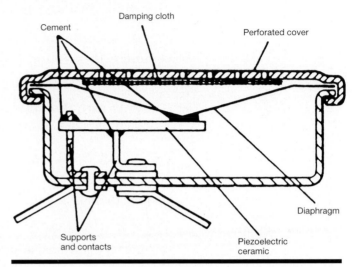

Figure 7-2. *Schematic of a piezoelectric microphone.* (*Reprinted with permission of Morgan Matroc Ceramic Division, Bedford, OH.*)

The diaphragm in a microphone amplifies the vibrations from the sounds entering the microphone and transmits them to the piezoelectric ceramic. Each combination of sounds produces a unique pattern of vibrations that push against the ceramic. The piezoelectric ceramic converts these tiny pressures into a unique electrical signal that matches the unique pattern of vibrations. This electrical signal can then be stored or transmitted to a speaker where it's converted back to the original, unique pattern of sound waves by means of a piezoelectric ceramic or a magnetic device (using a ceramic magnet). Piezoelectric ceramics are especially good at producing higher-pitched sounds, so they're usually used for the "tweeter" speakers in your stereo, whereas ceramic magnets work best for the lower-pitched "woofer" speakers.

All of the early piezoelectric materials were single crystals. In 1941, R.B. Gray obtained a patent on a piezoelectric polycrystalline ceramic called *barium titanate* (made up of atoms of barium, titanium, and oxygen). This was a major breakthrough as the polycrystalline ceramic was stronger and more durable than single crystals, cheaper to make, and not soluble in water like Rochelle salt. An early use of barium titanate was for fish finders. Improved polycrystalline piezoelectric ceramics were developed during the

1940s, culminating in the invention, in 1954, of *lead zirconate titanate* (PZT) by Bernard Jaffe. This material, made up of atoms of lead, zirconium, titanium, and oxygen, had especially strong piezoelectric behavior and opened the door for hundreds of new applications.

As scientists and engineers became more familiar with piezoelectric ceramics, they found a wide range of imaginative ways to use these remarkable materials.

• As with the microphone, they could use piezoelectric ceramics to detect vibrations, even those as minute as sound waves.

• They could apply an alternating current, such as the electricity you get every day when you plug in your TV or toaster, to a piezoelectric ceramic to make it vibrate.

• If they applied a direct current, such as from a battery, they obtained a controlled amount of mechanical deflection (movement, shape change) of the ceramic. The amount of deflection could be controlled from millionths of an inch to tenths of an inch by the amount of voltage applied, the thickness of the piezoelectric ceramic, and the number of layers of ceramic that could be built into a stack.

• They observed that sometimes the electrical input forced the piezoelectric ceramic to vibrate at its natural frequency (like a rubber band vibrates when you pluck it) and either emit sound, like the vibrating string of a guitar, or create movement that could be used as a measurement of time, like in a quartz watch.

• They learned ways to apply a controlled amount of pressure or a sharp blow to the piezoelectric ceramic and produce a high-voltage spark that could ignite a fuel, as an alternative to a spark plug.

• Much more recently, they learned that piezoelectric ceramics could convert unwanted vibration into electrical energy that can be removed, as heat as in the "smart skis" marketed by the K2 company.

The rest of this chapter presents many different uses for piezoelectric ceramics that rely on these different ways piezoelectricity works.

UNDERWATER USES FOR PIEZOELECTRIC CERAMICS

The ability of a piezoelectric ceramic to vibrate when the electricity is turned on or, conversely, to detect vibrations and convert them into electrical signals, makes possible many underwater applications such as hydrophones, sonar, devices for tracking schools of fish, precision speedometers for boats, and even equipment for mapping the bottom of the oceans. Let's look at some of these underwater applications and how piezoelectric ceramics make them possible.

Hydrophones

An early use for piezoelectrics was in hydrophones, which were important during the Second World War for detecting enemy submarines and monitoring other ships from submarines. Remember the old submarine movies? A submarine would have all its engines off, and the crew would be listening electronically for sounds made by other submarines or ships. Hydrophones were the devices that made this clandestine listening possible.

UNDERWATER USES FOR PIEZOELECTRIC CERAMICS

Hydrophones

Sonar

Decoys and homing beacons

Boat speedometers

Fish finders

Underwater explosives detectors

Ocean floor mapping devices

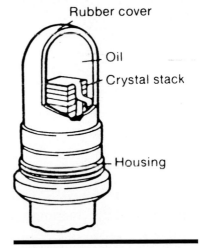

Figure 7-3. Schematic of a hydrophone.
(Reprinted with permission of Morgan Matroc Ceramic Division, Bedford, OH.)

As shown in Figure 7-3, a hydrophone consisted of a stack of piezoelectric crystal slices (usually lithium sulfate) immersed in oil inside a sealed rubber-and-metal housing. Being water soluble, the lithium sulfate couldn't be immersed directly in water. Sound waves from the engine of another submarine or a surface ship would travel through the water and strike the hydrophone. The piezoelectric crystals detected the pressure of these sound waves and gave off electrical signals that could be amplified and shown visually on the screen of an oscilloscope (similar to a small television screen), just as portrayed in the submarine movies. The way the signal looked and acted let the submarine crew estimate the type of ship they'd spotted, its direction from them, and the direction it was moving, although not its distance from them.

Sonar

Sonar, which stands for "sound navigation and ranging" was an improvement over the hydrophone. It was developed to detect, as well as to determine the distance of, underwater objects such as another submarine, even with its engines off, or the hull of a surface ship. Sonar can be compared to the navigation technique of a bat or to the radar gun used by police to catch speeding motorists. Certain types of waves sent out from a device travel through the air (in the case of the bat or radar gun) or water (in the case of sonar) until they hit a solid object. Some of these waves bounce off the object and come back, to be detected by the sender.

In the case of sonar, sound waves are emitted. A sound wave is produced by applying a pulsed electrical input to a piece of piezoelectric ceramic. Each electrical pulse causes the ceramic piece to distort, or bend, which produces a sound (acoustic) wave in the water. A piezoelectric ceramic used in this way is called a *transducer*. Each wave spreads out and travels through the water, just like a wave spreads out and travels away from a pebble that you drop into a pond. When the waves run into a solid object, they reverse direction and head back to the transducer. Now the transducer acts like a hydrophone and converts the pressure from the waves back into electrical pulses. Because the ship's crew knows the speed of sound in water, they can precisely calculate the distance from them to the object. Sonar works as the eyes of the submarine, just as a pulse-echo system replaces the eyes of a bat in total darkness.

Decoys

Because sonar gives us a way to detect and track both underwater and surface ships, it has great military value. Of course, every military device seems to have a counter-device designed to defeat it. How do you defeat a piezoelectric device? With another piezoelectric device. A piezoelectric "pinger" is spherical and sends out an acoustic signal in all directions. Pingers can be released by a submarine or ship to act as decoys to send false signals to enemy ships. A pinger also can be used as a "homing" or directional beacon.

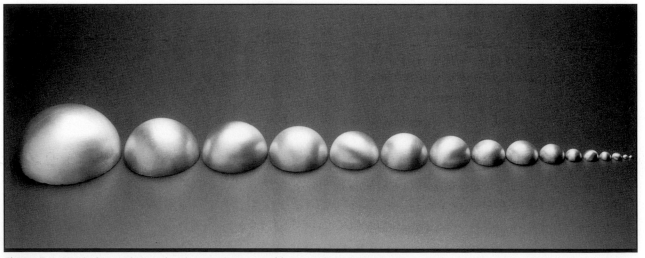

Figure 7-4. *Hemispheres of piezoelectric ceramic to assemble into spherical pingers* (Photo courtesy of Edo Corporation Piezoelectric Products, Salt Lake City, UT.)

Boat Speedometers

A device similar to sonar, except with two stacks of piezoelectric ceramic slices oriented perpendicular to each other, is used as a speedometer for boats and ships. Called a *doppler sonar velocity system* and mounted on the bottom of the ship, this device sends ultrasonic vibrations to the bottom of the body of water and uses the reflections to calculate the speed of the boat. It can even correct for ocean currents. Accurately measuring the speed of an ocean-going vessel always has been a challenge. The

Figure 7-5. Doppler sonar velocity system. Measuring approximately 6 inches in diameter, the system can measure the speed of a ship to a precision of 0.2 inches (0.5 centimeters) per second. (Photo courtesy of Edo Corporation Piezoelectric Products, Salt Lake City, UT.)

piezoelectric device is a big improvement over the previous system: dragging a rope with knots tied at regular spaces and counting the number of knots that pass a floating object during a certain time.

Fish Finders

Another category of piezoelectric ceramic devices that operates in the same fashion as sonar is used for finding and tracking schools of fish. Fish finders have been in use since the late 1940s and have dramatically increased the efficiency of the world fishing industry, unfortunately even resulting in overfishing in some areas.

Underwater Explosives Detectors

Explosive mines are a major threat to ships during wartime. Sonar can detect a mine, but conventional sonar aims a beam in only one direction at a time. When large clusters of sonar-like transducers are assembled in a circular array, as shown in Figure 7-6, they can search in all directions. The whole array is about 5 feet (1.5 meters) high and over 10 feet (3 meters) in diameter and contains hundreds of piezoelectric transducer stacks. The array is suspended underwater from a ship. Reflections from any object encountered by sound waves sent out by the array of transducers are

Figure 7-6.(right) *Array of sonar transducers for searching for underwater explosives.* (Photo courtesy of Edo Corporation Piezoelectric Products, Salt Lake City, UT.)

Figure 7-7. *Sequence used to assemble a single transducer stack for the array shown in Figure 7-6.* (Photo courtesy of Edo Corporation Piezoelectric Products, Salt Lake City, UT.)

detected by the array and analyzed by a computer to determine the position and size of the object and whether it is stationary or moving. Minesweepers can then move in to check suspicious objects.

Ocean-Floor Mapping Devices

Piezoelectric transducers are also important tools for mapping the ocean floor and searching for oil-bearing formations beneath the sea. Piezoelectric transducers bounce sound waves off the ocean floor and record the reflected waves to allow calculations of the water depth. Mapping is repeated over a wide area, and the results are combined to construct a three-dimensional image of the ocean floor. To search rock

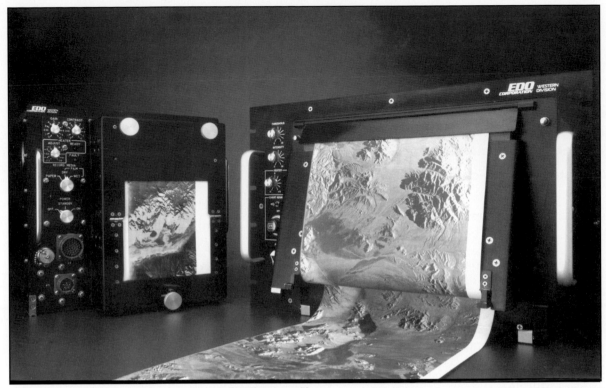

Figure 7-8. *Ocean-floor contour map from a piezoelectric profiler.* *(Photo courtesy of Edo Corporation Piezoelectric Products, Salt Lake City, UT.)*

formations underneath the ocean floor for oil, much stronger sound waves than can be produced by a piezoelectric transducer are necessary. These stronger waves are produced by air cannons set off at the surface of the ocean. The waves still bounce, or reflect, off the ocean floor, but they also continue into the rock formations and send back reflections from each rock layer, to be detected by piezoelectric transducers.

INDUSTRIAL USES FOR PIEZOELECTRIC CERAMICS

Piezoelectric ceramic materials have become very important for a wide variety of uses in industry. Some of these uses include inspecting solid materials such as steel for hidden internal defects, cleaning oil and debris off as-fabricated parts before they're assembled into a product, mixing difficult-to-mix liquids and even ceramic slurries, and precision controlling industrial machines.

Nondestructive Inspection Devices

Suppose you're a construction engineer and you've just bought some 8 inch thick beams of steel to build a bridge. From your plans for the bridge, you know that any pores, cracks, or other defects larger than one-quarter of an inch will lead to disaster. You can inspect the surface of the beams visually, but how do you look inside the beams without cutting into them? One effective approach that lets you "see" inside an object without tearing it apart is called *ultrasonic nondestructive inspection* (ultrasonic NDI).

INDUSTRIAL USES FOR PIEZOELECTRIC CERAMICS

Nondestructive inspection devices

Ultrasonic cleaners

Mixers

Emulsion preparation equipment

Actuators for industrial machines

Ultrasonic machining equipment

Motors

Transformers

Ultrasonic NDI works just like sonar. A ceramic piezoelectric transducer is placed against the surface of the object—in this case, our steel beam—and caused to vibrate. The sound waves produced by the vibrations travel through the steel beam and are detected by another piezoelectric transducer. If the steel has any internal defects, they will interrupt the sound waves, whereas steel without flaws will allow the sound waves to pass through. Scanning the transducer back and forth allows the whole steel beam to be inspected. Ultrasonic NDI is used extensively to assure the quality of materials for construction and for many industrial applications. It has even been used to continuously monitor the thickness and quality of paper coming out of a papermaking machine.

Ultrasonic Cleaners

Nondestructive inspection is only one of numerous important industrial uses for piezoelectric ceramics. Have you gone to the jewelers lately to have your rings or watch cleaned? Ultrasonic cleaning is a familiar piezoelectric task. High-intensity ultrasonic waves reverberating through a container of water or other fluid cause bubbles to form at the surface of anything immersed in the fluid. The formation of these bubbles produces a vigorous cleaning action. Ultrasonic cleaning is standard for many industrial processes and now is often even used in our homes for cleaning such items as jewelry and dentures.

Equipment for Mixing and for Making Emulsions

Often two liquids, or a liquid plus solid particles (such as a ceramic in a slurry, or a slip), must be mixed to the best uniformity possible. Ultrasonic waves from a ceramic transducer, such as in an ultrasonic cleaner, turn out to be one of the most effective ways to obtain a very uniform mixture.

The ultrasonic concept has been effective for helping manufacturers form emulsions, which are stable mixtures of two liquids that are not soluble in each other such as most oils and water. Normally, a mixture of water and oil will separate after standing for a while, and the oil will rise to the surface; that's why it's necessary to shake salad dressing before serving it. But a true emulsion is such a fine mixture of the oil and water that they do not separate. A good example is a hand lotion or some shampoos. Ultrasonic vibrations and bubble generation help to form a true emulsion by breaking down the oil droplets into such tiny globules in the water that they don't re-form into larger globules and float to the surface.

Actuators and Positioners

A different industrial use for piezoelectric ceramics is as an *actuator* or *positioner,* to precisely control the movement or position of a manufacturing machine or even the end product. When a piezoelectric ceramic is used as an actuator, it isn't vibrated. Instead, a direct current of electricity is applied, causing the piezoelectric ceramic to change shape. How much the piezoelectric ceramic moves, or deflects, depends on the size and shape of the ceramic piece or stack and on the amount of electricity applied. Very tiny movements, down to millionths of an inch—much more precise than those that can be achieved by conventional actuators—can be created in this way. Such a degree of control is necessary in some machining operations, in positioning a laser beam, and in controlling robotic movements during precision manufacturing.

Direct-current, rigid-motion piezoactuators are used in many other ways in industry and in the products of industry. One interesting use is for tiny microswitches and pressure-sensitive valves, often as safety features so that a piece of equipment will shut down if it begins to operate outside its normal safe range. Another category of uses for piezoactuators is in measurement equipment, to make sure that manufactured items are the right shape and size to fit into precision products. The piezoactuator allows the equipment to make the tiny movements necessary to measure down to millionths of an inch, whereas mechanical devices such as calipers can only measure accurately to thousandths of an inch.

Recent advances in piezoactuator materials and designs have increased the range of movement from millionths of an inch to tenths of an inch, which has opened up many new application opportunities. Hundreds of patents are applied for each year for innovative new actuator applications in many different fields. The technology has even been applied to cameras, in the form of a piezoelectric autofocus actuator motor.

Ultrasonic Machining Equipment

Cutting or machining a complex shape into wood or a metal is fairly easy; but it's difficult and expensive to cut the shape into a piece of rock or ceramic, because the ceramic is so hard and resistant even to scratching. An ultrasonic method has been invented that's almost like cutting out cookies in your kitchen. First, a metal tool is made that looks like a cookie cutter but has the outline of the shaped part you want to make. The tool is attached to a piezoelectric transducer and caused to vibrate. A liquid containing tiny particles of a hard ceramic powder, such as diamond or silicon carbide, is flowed over the surface of the tool. When the "cookie cutter" tool is brought into contact with the part, the ceramic particles in the liquid vibrate between the tool and the part and slowly cut a hole through the part in the shape of the tool.

Transformers

Different types of electrical equipment in industry run on different amounts of electricity, and so the electricity coming into a manufacturing plant must be either stepped up (increased) or stepped down (decreased) as needed. This is done with a *transformer,* which is usually made with expensive coils of copper wire. A transformer that can step up or step down a dc electrical source can also be made with piezoelectric ceramics. The transformer is fabricated by bonding thin strips or stacks of piezoelectric ceramics together in a special way, so that electrically stimulating one stack, which causes it to deform, will cause even greater deformation in the second stack. The second stack then gives off a higher voltage than the original voltage applied to the first stack. For example, one transformer was designed so that a 10 volt input produced a 400 volt output. Increasing the input to 30 volts resulted in an output of 2400 volts. The whole transformer measured only 2.75 by 0.39 by 0.09 inches (7 by 1 centimeter by 2.5 millimeters), much smaller and lighter than a comparable copper-coil transformer. Another advantage of the piezoelectric transformer is that it can perform in a magnetic environment that would interfere with a coil transformer.

PRESSURE AND VIBRATION DETECTION WITH PIEZOELECTRIC CERAMICS

Vibration is often a sign that a problem is present or that a piece of equipment is on the verge of failing. Piezoelectric ceramics are often used as sensors to detect the onset or degree of vibration. The vibrations are tiny pressure pulses that cause the piezoelectric ceramic to generate an electrical signal. The electrical signal can be monitored continuously by a simple microprocessor, or by a computer programmed to set off an alarm or activate a switch to shut off the equipment if a critical level of vibration is detected. Such sensors are used on aircraft, in motor vehicles, and for various kinds of industrial equipment. They allow the user to shut down the equipment for maintenance before a much-more-expensive and dangerous catastrophic failure occurs.

One use of piezoelectric vibration detection that we all encounter if we buy new tires is a "spin" wheel balancer. The mechanic mounts the tire on the wheel and places it on a machine that spins the tire at about the same speed the tire would be moving on the highway. If the tire and wheel are balanced (if the weight's distributed evenly around the axle), no

vibrations show up. If the tire plus wheel are out of balance, the machine will shimmy and shake, just like your car would on the highway, and the piezoelectric sensor will detect the vibration. The mechanic then attaches lead weights at points around the rim of the wheel until the sensor measures that the vibration is reduced to acceptable limits. A tire that isn't properly balanced will vibrate against the road and wear irregularly, reducing tire life and causing a safety hazard.

Another safety application for piezoelectric sensors is in automotive airbags. The sensor can detect an impact, and it can be set to discriminate whether the impact is severe enough to deploy the airbag.

An area of great interest currently is *smart contact sensors*. These sensors "feel" the pressure of contact and send an electrical message to a computer chip that will cause some sort of action. You can probably imagine all kinds of uses for such sensors in robots, industrial machines, artificial hands, and virtual-reality devices.

Figure 7-9. *Low-cost, miniature sensors for detecting impact or vibration.* (Photo courtesy of Edo Corporation Piezoelectric Products, Salt Lake City, UT.)

VIBRATION PREVENTION

Piezoelectric ceramics not only help detect vibration where it isn't wanted; they also can prevent it. The most important such application is in vibration-free tables for the photolithography steps used in making

USES FOR VIBRATION PREVENTION BY PIEZOELECTRIC CERAMICS

Vibration-isolation tables

"Smart" skis

Aircraft structures requiring vibration reduction

Baseball bats

silicon-microchip integrated circuits. Other uses include reducing the vibrations in aircraft structures, skis, and baseball bats.

Active Vibration Control for Microchip Production

Vibration is a big problem in the fabrication of integrated circuits. As we discussed in Chapter 6, the photolithography steps must produce details only about 20 millionths of an inch (500 nanometers) across. The slightest vibration will blur the pattern being projected through the photolithography lens and result in microchips that don't work. Vibrations from cars and trucks passing by, airplanes flying overhead, workers walking in the room, and all other sources must be completely eliminated. The table on which the photolithography is conducted has two sets of piezoelectric devices mounted in its legs. One set detects any vibrations trying to travel up the legs of the table and sends a signal to a computer that controls a second set of piezoelectric actuators farther up the legs of the table. The computer calculates automatically just the right amount of electricity to apply to each actuator on each table leg to produce just the right amount of pressure to cancel out and stop the vibrations. Each actuator can apply forces up to about 1000 pounds (455 kilograms), which can handle just about any level of vibrations short of a major earthquake. This use of piezoelectric ceramics is referred to as *active vibration control.*

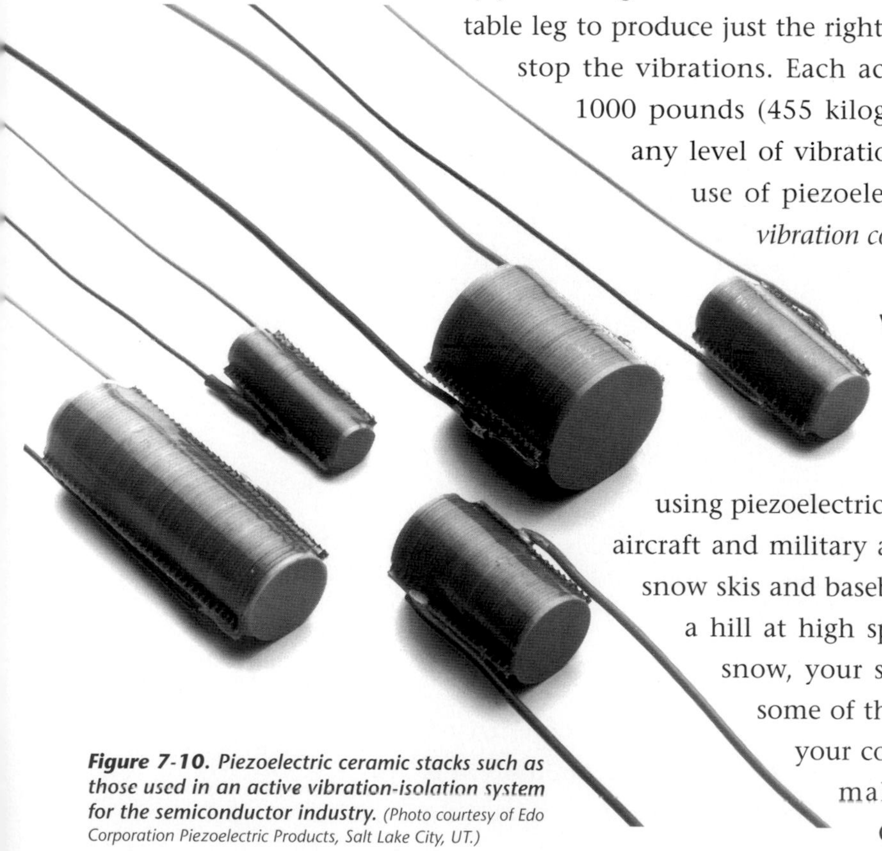

Figure 7-10. ***Piezoelectric ceramic stacks such as those used in an active vibration-isolation system for the semiconductor industry.*** *(Photo courtesy of Edo Corporation Piezoelectric Products, Salt Lake City, UT.)*

Vibration Reduction in Aircraft, Skis, and Baseball Bats

Another vibration-control system using piezoelectric ceramics was developed originally for aircraft and military applications but has been adapted for snow skis and baseball bats. When you're zooming down a hill at high speed on your skis, especially on hard snow, your skis start to vibrate (*chatter*) and lose some of their contact with the snow, decreasing your control of the skis. The K2 Company (a maker of ski equipment) and Active Control eXperts (a designer and maker of piezoelectric devices) have come up with a piezoelectric solution. They build a piezoelectric ceramic module, shown in Figure 7-11, into each ski. A module is made up of three pieces of piezoelectric ceramic, each about the size and shape of a small snack cracker, embedded along with some electrical wiring in a sheet of plastic. When the skis start to vibrate, the piezoelectric ceramics in the module convert some of the vibration to electricity, which flows into the electrical wires. These wires are made of a material that isn't a particularly good conductor of electricity (it has some electrical resistance, like the filament in a light bulb), so the electricity is converted to heat. This conversion of vibration to electricity to heat essentially dampens, or erases, close to 40 percent of the vibration, allowing your skis to hug the snow and you to maintain an increased level of control. These skis even have a tiny, ceramic light-emitting *diode* built into the piezoelectric module, which lights up when electricity is flowing, so that you can see when your "smart skis" are working!

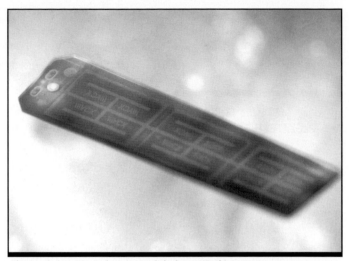

Figure 7-11. *Piezoelectric module for a K2 ski.* (Photo courtesy of Active Control eXperts (ACX), Inc., Cambridge, MA.)

These "smart" skis are clever. Who knows what engineers will come up with next? Maybe they'll put the electrical wiring in your ski boots and have a plug-in that connects your boots to the piezoelectric module in your skis, to help keep your feet warm while you're skiing.

The same active vibration-control concept used for the skis works for baseball bats. You've probably hit a baseball or softball and felt the sting of the force and vibration on your hands. A piezoelectric module built into the bat gets rid of a lot of the sting. Before long, piezoelectric modules may be built into tennis rackets, golf clubs, and maybe even jogging shoes!

Figure 7-12. *Skier wearing K2 "smart skis."* (Photo courtesy of ACX, Inc., Cambridge, MA.)

OTHER USES FOR PIEZOELECTRIC CERAMICS

There are still many other ways that piezoelectric ceramics are used, but we'll only talk about a few more in this chapter: the quartz watch, devices in which the ceramic gives off a sound that we can hear, lighters and igniters, and actuators to counteract temperature distortions in telescope mirrors. In the next chapter, we'll talk about some very important medical applications of piezoelectric ceramics, along with other surprising uses for ceramics in the field of medicine.

Quartz Watches

Perhaps the most common piezoelectric application is in the quartz watch. A quartz watch contains a small slice of quartz crystal, cut in just the right way to make it piezoelectric. Connected to the battery in the watch, the slice of quartz vibrates at a constant frequency (thousands of vibrations per second). Each of these vibrations is equal to a tiny unit of time. Using these vibrations, the quartz watch accurately measures time and is much more dependable than old-fashioned mechanical watches, with their springs, levers, gears, and bearings.

Noisemakers and Beautiful Music

Remember our little experiment earlier in this chapter, when you felt the vibrations on your Adam's apple as you hummed high and low? Sounds you hear are simply vibrations. Vibrations with the peaks of the waves close together (short wavelength, high frequency) are high pitched, like the sounds made by an opera singer reaching high C or the notes that result from the keys on the far right of a piano keyboard. Vibrations with the peaks of the waves farther apart are lower pitched, like those made by the cello in the orchestra rather than the violin. How do we get different pitches of sound? In stringed musical instruments, the sound is controlled by the diameter and length of the string. Small diameter, short, tightly strung strings produce high notes. Strings that are bigger in diameter, longer, and strung less tightly produce lower notes. In bells, chimes, or tuning forks, the pitch is controlled by the size of the instrument itself. For example, if you've ever listened to a bell choir, you know larger bells produce lower sounds.

Engineers have learned to produce a complete range of sounds with piezoelectric ceramics by carefully controlling the size and shape of the ceramic piece and the way that electricity is applied to it. In one case,

---

OTHER USES FOR PIEZOELECTRIC CERAMICS

Quartz watches

Buzzers, alarms, and pagers

Musical greeting cards

Musical instruments

Lighters and igniters

Telescope-mirror correction actuators

---

they've designed the ceramic to give off the jarring sound of an alarm clock, to make sure we don't roll over and go back to sleep, or the harsh sound of a smoke alarm, burglar alarm, or seatbelt buzzer. In another case, they've designed the ceramic to produce a clear, pure musical sound that can rival that from the best musical instruments. By making a separate ceramic piece for each note in the musical scale and having a separate electrical switch for each, designers can create the sound of a musical instrument or a whole orchestra. This idea has even been miniaturized to make possible musical greeting cards and toys that require only a small battery.

Lighters and Igniters

The first piezoelectric effect that was seen by cave dwellers—the spark given off when a piece of quartz was struck with a rock—is now used for lighters and igniters. Many lighters, for example, contain a burnable fuel that's ignited with a tiny piece of piezoelectric ceramic measuring only about 1 by 1 by 3 millimeters. When you "flick your Bic," a tiny metal piston is pushed against a spring, away from the ceramic piece, then released and forced by the spring to strike the piezoelectric ceramic with a sharp blow. The blow causes the piezoelectric ceramic to give off a spark that ignites the lighter fluid. This same concept is used for charcoal-grill lighters. The lighter produces a small impact and pressure that results in a small spark.

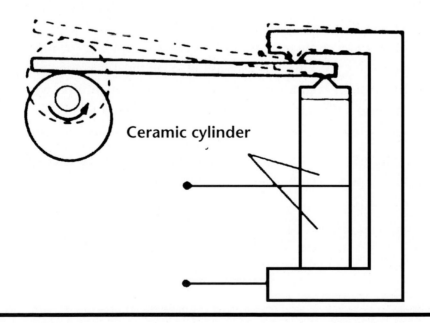

Figure 7-13. Schematic of a spark generator. *(Reprinted with permission of Morgan Matroc Ceramic Division, Bedford, OH.)*

Using a larger piece of ceramic and higher pressure can produce a very large, high-voltage spark. The spark generator shown in Figure 7-13, designed as an alternative to a spark plug in a gasoline-burning engine, produced a 21,000 volt spark by squeezing a couple of half-inch-square pieces of piezoelectric ceramic together with a pressure pulse of about 7000 pounds (3180 kilograms).

Actuators for Correcting Temperature Distortions in Telescope Mirrors

In Chapter 5 you learned that temperature variations in large telescope mirrors cause distortions that can blur the image the astronomer is trying to observe. New, thin mirrors have been designed with an array of piezoactuators attached to the back side. These actuators can be computer controlled to slightly change the shape of the mirror, counteracting any distortions caused in the mirror by temperature changes, and thus improving the quality of pictures that can be taken of distant galaxies.

OVERVIEW: PIEZOELECTRIC CERAMICS TRANSFORMING OUR LIVES

Piezoelectricity seems stranger than truth. Ceramics that can convert from electricity to pressure and vice versa have introduced the world to microphones and phonographs, to sonar and fish finders, and to a nondestructive way of seeing inside solid objects. They help us make integrated circuits, act as the timekeepers in our watches, and give us advanced warning before costly machines self-destruct. In spite of the many uses discovered so far, inventors are just starting to tap the true commercial potential of this magical technology. In the not-too-distant future, we may have smart sensors that give most of the products we use a "sense of feel" and the ability to make preprogrammed decisions controlled by a silicon chip.

The next two chapters discuss the enormous impact of ceramics (including piezoelectric ceramics) in the fields of medicine and automobiles.

Medical Miracles

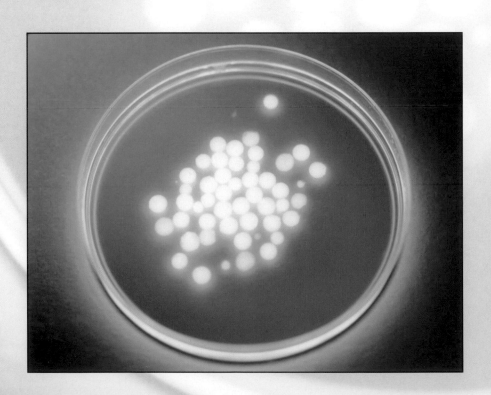

CHAPTER EIGHT

*I*n the last few chapters we've explored different properties of ceramics—such as strength, electrical characteristics, and optical behavior—and then discussed some of the useful products these properties make possible. Let's take a different approach in this chapter. Let's pick a single field of application—medicine—and look at the different ways ceramics are used. You'll be surprised at the breadth of contributions by ceramic materials.

Probably the earliest medical use of ceramic materials was as the glass lenses in eye glasses, a use that goes back to at least the 12th century. This was followed much later, in 1774, by the first successful use of ceramics (porcelain) as a replacement for teeth. Most of the applications that really can be classified as miraculous, however, have been developed during the last 40 years. These modern medical innovations can be divided into three categories: (1) replacement and repair, (2) diagnosis, and (3) treatment/therapy.

PRODUCTS AND USES

Tooth and jaw repair
Joint implants
Bone repair
Middle-ear implants
Heart valves
Eye-lens replacement
Artificial eyes
Prosthetics

CERAMICS FOR REPLACEMENT AND REPAIR

One of the most exciting areas of medical advancement during our generation has been the repair or complete replacement of damaged parts of our bodies. For a replacement to be successful, the material used must be able to perform the same function as the original body part and also must be compatible with the surrounding tissue. These sound like fairly simple requirements but actually are quite challenging, especially with regard to compatibility. Many materials are toxic to tissue. Most others that aren't toxic instead are attacked by the body and either destroyed or encapsulated in special fibrous tissue (similar to scar tissue) that our body builds up in an effort to isolate the foreign part that has been implanted. Some

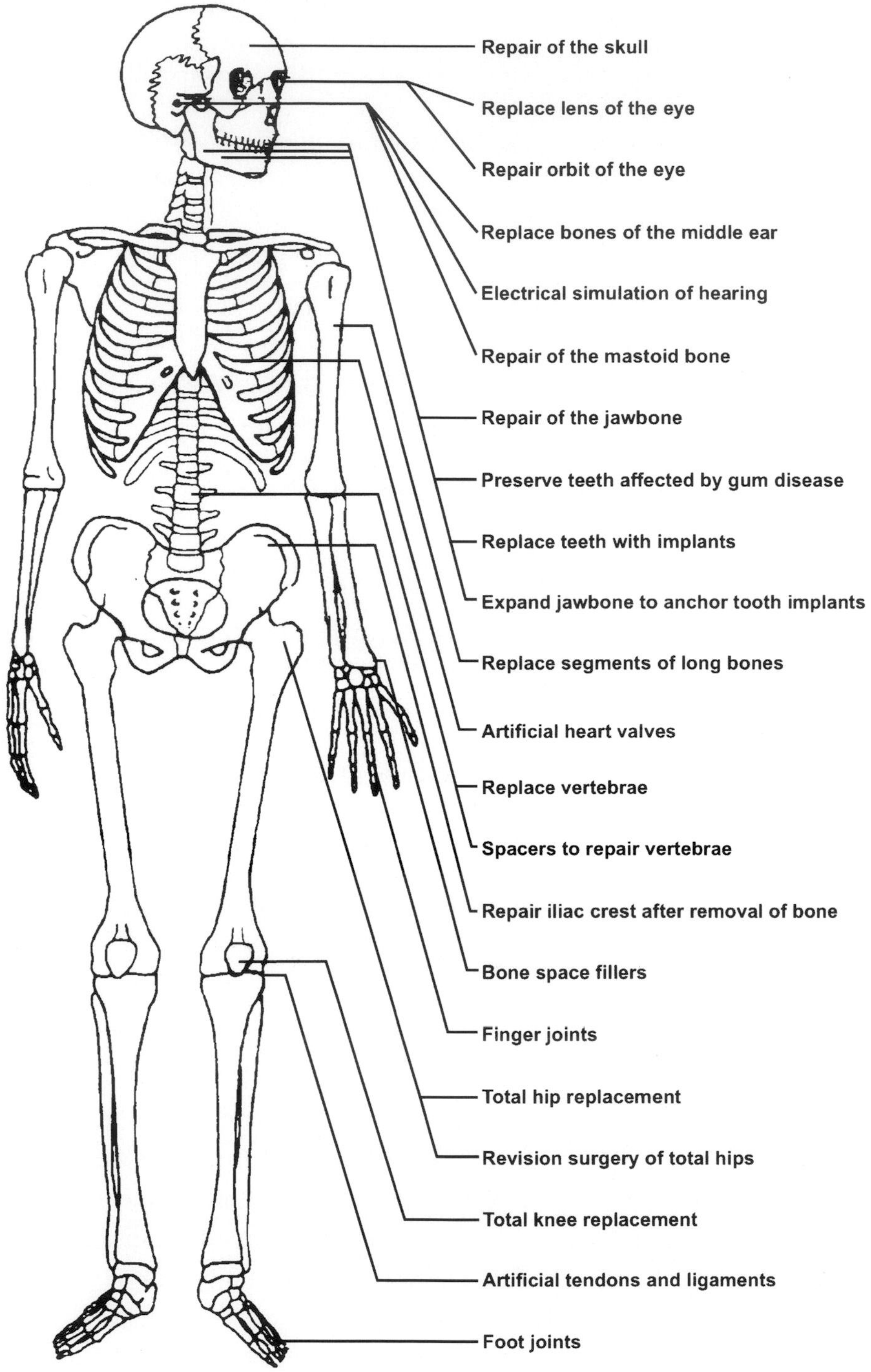
Repair of the skull
Replace lens of the eye
Repair orbit of the eye
Replace bones of the middle ear
Electrical simulation of hearing
Repair of the mastoid bone
Repair of the jawbone
Preserve teeth affected by gum disease
Replace teeth with implants
Expand jawbone to anchor tooth implants
Replace segments of long bones
Artificial heart valves
Replace vertebrae
Spacers to repair vertebrae
Repair iliac crest after removal of bone
Bone space fillers
Finger joints
Total hip replacement
Revision surgery of total hips
Total knee replacement
Artificial tendons and ligaments
Foot joints

Figure 8-1. *The Bioceramic Man. Ceramics are used from head to toe as implants, repairs, and restorations. (Courtesy of Professor Larry Hench, Imperial College, London.)*

crystalline ceramics and glass, however, mimic the composition of bone (which is mostly ceramic) and have proved to be the most body-compatible materials found so far. These biocompatible ceramic materials are referred to as *bioceramics*.

The story behind early bioceramics work in the United States is interesting. A young ceramic engineer named Larry Hench, at the University of Florida, was working on semiconductor switches for nuclear weapons. On his way to an Army materials conference in 1967, he sat next to a colonel who had just returned from Vietnam. The colonel complained about the thousands of soldiers who were having limbs amputated because their bodies were rejecting metal and plastic implants. He admonished researchers for working on tools of destruction rather than tools of healing.

Inspired to action, Hench proceeded to win a grant from the U.S. Army in 1969 to explore ceramic materials for their compatibility with tissue. He started his studies with rats, which respond to medical procedures in much the same way that people do. Within two months he had identified a glass composition (based on calcium and phosphorus, the major constituents of bone) that bonded to bone and tissue better than any previous material. This new material became known as *Bioglass*.

Bioceramics are now successfully implanted in our bodies as solid parts, porous parts, and coatings. Those that actually bond with our bone and tissue are called *bioactive*. Bioglass and polycrystalline ceramics with similar composition to Bioglass, such as a ceramic called *hydroxyapatite,* are the most successful bioactive ceramics. Some of these compositions are actually slowly absorbed by our body as they are replaced with natural bone and tissue. Other ceramic materials, such as alumina and zirconia, don't bond to our bone and tissue but are neither toxic nor attacked by our body and rejected. These inert ceramics also have become important as implants, especially for the replacement of damaged or worn out joints.

Dental Ceramics

The first recorded attempt to use ceramics as replacements for teeth was in about 1774, in Paris, France, by a dentist, Nicholas Dubois de Chémant. Before that time, many other materials had been used to replace teeth, including ivory, bone, wood, animal teeth, and even teeth extracted from human donors. These prior tooth replacements weren't particularly effective, because they quickly became stained and invariably developed a bad odor. Dubois de Chémant saw pottery of the new porcelain that Europeans had just learned how to make and marveled that the appearance was so similar to that of natural teeth. He collaborated with Josiah Wedgwood, who, you may recall from Chapter 3, was prominent in establishing porcelain manufacturing in Britain. The resulting porcelain teeth were a big improvement over the previous materials.

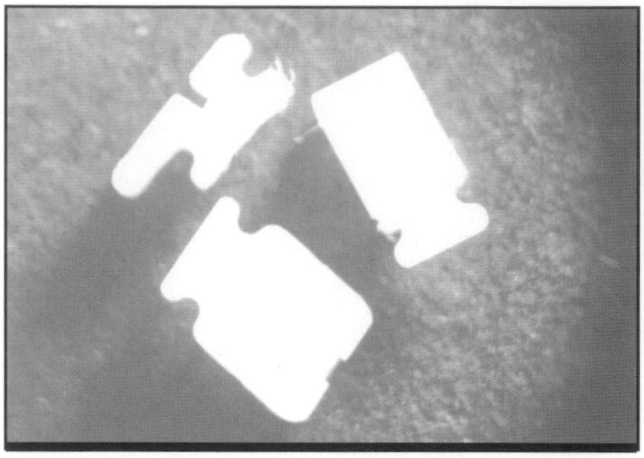

Figure 8-2. *Translucent alumina orthodontic brackets.*
(Fabricated by Ceradyne, Inc., Costa Mesa, CA. Photo by D. Richerson.)

Since 1774, and especially during the past 40 years, dental ceramics have been refined to duplicate the color and translucency of natural teeth and to provide higher strength. Not only have ceramics been used for tooth replacement and for dentures, but new restoration techniques, such as veneers, inlays, and crowns, have been developed. A veneer consists of a thin layer of ceramic, usually bonded to the visible surface of front teeth. An inlay is a filling. A ceramic inlay provides a natural-colored alternative to repairing a cavity with a silver-tin-mercury filling. A crown is a hollow ceramic cap, or shell, that completely covers the outer surface of a tooth.

Figure 8-3. *Comparison of metal (left) and ceramic orthodontic brackets.* *(Photos courtesy of 3M Unitek Corporation, Monrovia, CA.)*

Ceramics also are gaining importance in the field of orthodontics (the aligning of teeth that are crooked and out of place). To align teeth, an orthodontist cements a bracket onto each tooth and stretches a metal wire across the brackets to apply a steady pressure to push the teeth into place. The brackets traditionally have been metal, but the bright metal is so readily visible that some people are reluctant or embarrassed to have orthodontic braces. Initial attempts to replace the metal brackets with ceramics weren't successful. The stress caused when the orthodontist cinched up the wire was too great, and the ceramic usually broke. However, during the early 1980s, sapphire brackets were successfully introduced, and soon translucent polycrystalline alumina brackets followed. These ceramic braces, which are much less visible than metal braces, are now highly popular.

Hip and Other Joint Implants

If you've ever suffered damage or deterioration of your joints, especially the hip joint, you understand how much disability and suffering such problems can cause. Hip replacements were pioneered using a metal rod (stem) with a metal ball on one end. The stem was forced down the center of the leg bone (femur) and glued in place. The ball then fit into a plastic (usually the common polymer polyethylene) cap that was glued into the hip socket. However, these implants weren't considered a long term solution. Both the metal and plastic would slowly wear, and the resulting wear particles would cause irritation and inflammation. Within about 10 years, either the pain would become too great or the joint would freeze up and stop working; the implant would have to be replaced. Another problem also sometimes required replacement of the implant. The glue between the metal stem and the leg bone would fail, and the rod would come loose.

Ceramics have helped solve problems with hip replacements. Implants now last much longer than 10 years. Of the 950,000 hip replacements performed every year, about 250,000 are ceramic hip implants. The first problem-solving step was to replace the metal ball with transformation-toughened zirconium oxide, one of the new high-strength, high-toughness ceramics discussed in Chapter 5. Studies in Europe, Japan, and the United States in the 1980s showed that a ceramic ball was more body-compatible than metals and provided a harder, smoother surface that decreased wear. Another step was to cut grooves in the metal stem and coat them with a bioactive ceramic. This ceramic stimulated bone and tissue growth into the grooves and reduced the chance of the stem coming loose from the leg bone. Some new stems aren't even made of metal; instead, they're made of a composite of carbon fibers in a polymer matrix.

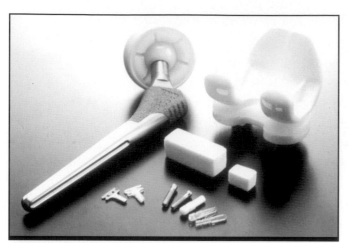

Figure 8-4. Hip replacement and other medical applications for ceramics. *(Photo courtesy of Kyocera.)*

This composite is a lot lighter and doesn't set off the metal detector at the airport! The latest innovation, which was just approved by the Food and Drug Administration (FDA), is a ceramic ball that can fit directly into the hip socket without requiring a polymer cap.

Major progress has been achieved during the last few years on other ceramic joint replacements, including the thumb, knee, big toe, and shoulder. Doctors still are learning about and perfecting these procedures, but such techniques are likely to become as common in the future as hip replacements are now.

Bone Restoration

Many medical conditions such as severe fractures can result in gaps or voids in your bone that prevent proper healing. Surgical procedures have been developed to take bone from another place and *graft* it in to fill gaps and replace damaged or diseased bone. Around 500,000 bone grafts were performed in 1994. Initial bone grafts were of two types: *autograft* and *allograft*. To autograft, the surgeon harvested bone from another place in the body to fill in for missing or damaged bone. Because autografts required a complex and expensive surgical operation, they often caused complications such as excess bleeding, chronic pain, deformity, and infection. Allografting used bone from a cadaver, which increased the risk of infection and also the potential that a patient's body would reject the graft.

A ceramic alternative to both the autograft and the allograft was introduced during the early 1980s and granted FDA approval. This ceramic material was developed by Interpore International of Irvine, California, and called Pro Osteon. Pro Osteon is made starting with natural coral (from the ocean), which has a pore structure very similar to that of bone. The coral is chemically modified in the laboratory, using heat and pressure, to change its chemical composition to the hydroxyapatite composition proven to be compatible with human tissue. The Pro Osteon is supplied to hospitals as small porous blocks that can easily be trimmed and shaped by a surgical team to fit the defect they're correcting in your bone. Within a week, your tissue and bone start to grow into the pores of the Pro Osteon. Within a couple of months, most of the implant has been replaced or interlaced with your own bone and tissue, and your bone is as strong and resilient as it was before the injury.

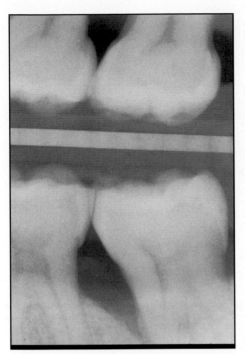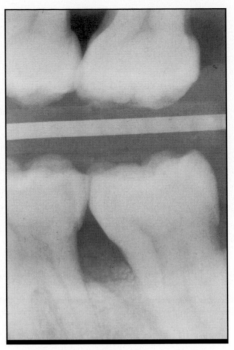

Figure 8-5. *Stimulation of bone growth around teeth by PereoGlas. The photograph on the left shows a large recess in the bone between the teeth. The photograph on the right shows that bone has grown into the recess after packing with particles of Bioglass (PereoGlas).* (Photos courtesy of U.S. Biomaterials, Alachua, FL.)

Middle-Ear Implants

Many people have hearing loss caused by disease or damage of the tiny bones (ossicles) that transmit sound vibrations from the ear drum to the inner ear. Plastic and metal replacements for these parts were tried without success; they either failed to bond to the ear drum or were attacked by the body and coated with fibrous tissue, which kept the sound waves from getting through. Then, U.S. Biomaterials, of Alachua, Florida, successfully used Bioglass developed by Larry Hench at the University of Florida. The Bioglass bonded to both soft and hard tissue and wasn't rejected by the patient's body. Thanks to a recent modification, surgical teams can now shape Bioglass implants during surgery to the precise contours of the ear.

Figure 8-6. *Four different sizes of Bioglass middle-ear implants.* (Photo courtesy of U.S. Biomaterials, Alachua, FL.)

Heart-Valve Implants

A common heart ailment is an aortic valve that doesn't close properly with each beat of the heart. This poorly functioning valve allows blood that should be flowing out of the heart to leak back in, reducing the amount of blood flowing from the heart with each beat. A ceramic mechanical heart valve has been successfully developed by the Medical Carbon Research Institute of Austin, Texas. The valve is made of a special carbon material called *pyrolytic carbon* and works a bit like a door with a spring attached. The valve (door) opens with each heart beat but closes between heart beats. Medical (pyrolytic) carbon also is effective for bone replacement.

Eye Repairs

We've used eye glasses for centuries to correct our vision from the outside, but glasses don't help if the inner parts of our eye are damaged or diseased. Researchers have made major strides in recent years toward finding ways to repair or replace eye parts. For example, the lens is the transparent, curved part of our eye that lets light in and focuses images onto the back (retina) of the eye. Sometimes the lens gets damaged, and sometimes it just gets clouded over—the condition called cataracts—as we get older. Until fairly recently, these conditions resulted in restricted vision or blindness. Now, surgeons can take out the damaged lens and replace it with a ceramic lens crafted of transparent aluminum oxide.

The laser has been an important innovation for eye repair. As we get older, our eyes change shape and our vision becomes blurred. This problem can be corrected by eye glasses, but many of us would prefer not to wear glasses. A new medical procedure uses a laser to reshape the lens of our eye and restore our perfect vision, without glasses. Other laser procedures can repair problems inside our eyes. For example, sometimes the retina of our eye comes loose (detached retina), which can cause blindness. Without having to cut the eye open, a doctor can shine a special laser through the lens and actually "weld" the detached retina back into place. Another inner eye problem is bleeding. If a tiny blood vessel inside our eye breaks, blood can flow onto the retina and permanently destroy the tiny vision receptors. If this condition is caught quickly enough, a laser can be used to seal the blood vessel to stop the bleeding, and even to clean up (burn off or vaporize) some of the blood that flowed onto the retina. These procedures are now widely available and have restored or retained the sight of many people.

Ceramics also are used as *orbital implants* (otherwise known as artificial eyes). Previously, artificial eyes simply filled a person's eye socket, but they weren't connected, or bonded, to surrounding tissue. If you had a "glass" eye, when you looked at something, your good eye would rotate to track the object, but your artificial eye would stay in one position. Now ceramic orbital implants actually bond to tissue in the eye socket, so that your artificial eye can follow your good eye to track an object.

Prosthetic Devices

Medical researchers have made amazing progress toward perfecting artificial legs, arms, and other prosthetic devices in recent years. Ceramics are playing an increasingly important role, in the form of high-strength fibers for reinforcing polymers. The fiber-reinforced composites have a very high strength-to-weight ratio, which means that very lightweight but strong prostheses can be designed and constructed from such materials. A new and exciting direction for development in this field holds great potential: integrating ceramic sensors into prostheses to provide a "sense of touch" and limit unwanted interaction with objects. This technology is still at an early stage of development, but piezoelectric sensors and other types of ceramic sensors look very promising.

USES OF CERAMICS IN MEDICAL DIAGNOSIS
CAT scanners
Ultrasound imaging equipment
Endoscopes
Optical instruments
Electronics
Computers
Communications networks

DIAGNOSTIC USES FOR CERAMICS

Ceramics are used extensively by doctors to help diagnose our ailments. Every hospital, clinic, and laboratory has virtually tons of glass containers to store blood and tissue samples and grow bacterial cultures; porcelain and alumina labware for chemical analyses; and electronics equipment—such as X-ray machines, spectroscopes, electrocardiographs, oscilloscopes, computers, and ultrasonic imagers—that rely on electrical and optical ceramics. These ceramic uses are fairly obvious and probably not surprising to you, but some special uses aren't so obvious, yet make possible some of the most important medical diagnostic tools: CAT scans, ultrasonic imaging, and endoscopes.

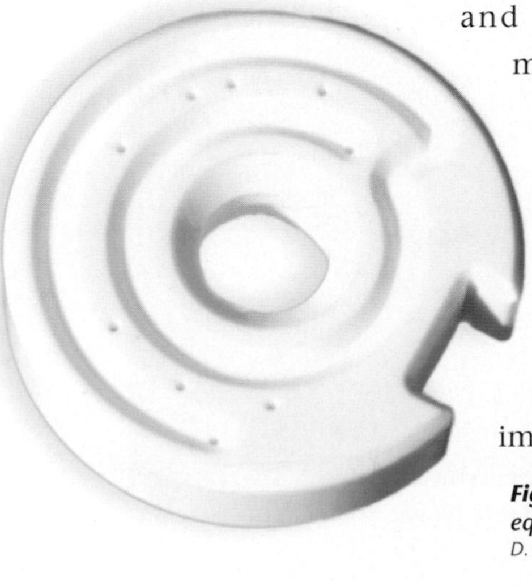

Figure 8-7. Alumina ceramic part used in blood-separation equipment. (Sample fabricated by Coors Ceramics, Golden, CO. Photo by D. Richerson.)

CAT Scans

The CAT scanner is a miraculous piece of equipment that allows doctors to see a remarkably clear and detailed view of our inner organs without having to cut into our bodies. Figure 8-8 shows some images taken with a CAT scanner. You can clearly see the brain, lungs, and other soft tissue organs as well as bones.

CAT stands for *computer-aided tomography* and is often referred to simply as CT. CT uses X rays, a powerful computer, and a special ceramic called a *scintillator*. You've probably had a regular X-ray examination, either in the doctor's or the dentist's office. The doctor places an enclosed sheet of photographic film on one side of your body (or teeth) and shoots

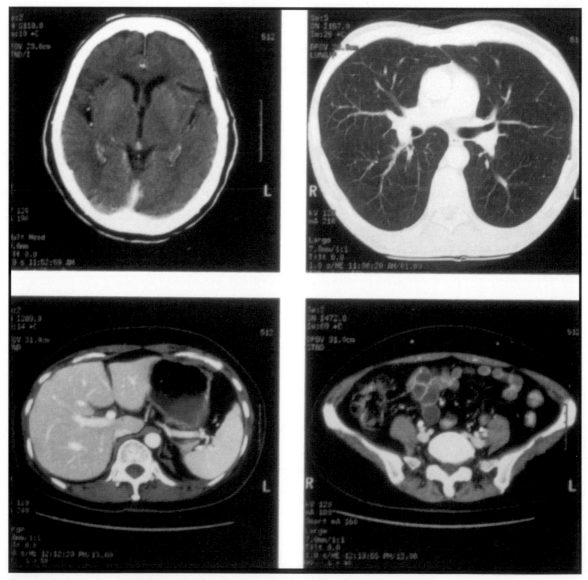

Figure 8-8. *Examples of CT images, showing the high resolution that can be achieved with ceramic scintillators.* (Photos courtesy of GE Medical Systems, Milwaukee, WI.)

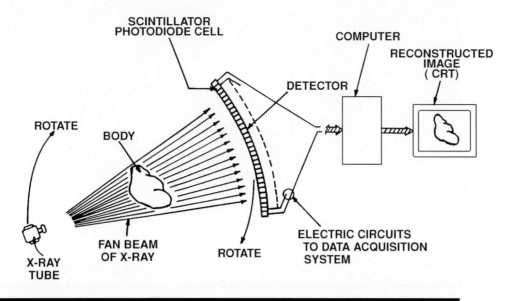

Figure 8-9. Schematic of a CT scan system. X rays pass through the body and are detected by an array of ceramic scintillators and analyzed by a computer to reconstruct an image of a cross-section view of the body. (Courtesy of General Electric Medical Systems, Milwaukee, WI.)

a broad beam of X rays through from the other side. The X rays that get through your body expose the film. Lots of X rays get through muscle and soft organs, but many fewer get through bone, so that the film is exposed in different amounts. This regular X-ray approach gives a pretty good image of your bones but not a very good image of your soft organs.

CT goes a step farther, as illustrated in Figure 8-9. In this case, a thin (rather than broad) beam of X rays is passed through your body and detected on the other side. Instead of film, though, a grid or miniature checkerboard of small slices of a ceramic scintillator is used as the detector (Figure 8-10). The scintillator ceramic is a special phosphor that glows when hit by X rays, just like the fluorescent rocks shown in Chapter 3 glowed under an ultraviolet light. The stronger the intensity of the X rays getting through your body, the brighter the glow of the scintillator. Attached to the back side of each tiny piece of ceramic scintillator is a *photodiode* that can convert the light into an electrical signal that varies according to the brightness of the glow. This electrical signal is fed as digital information into a computer.

The X-ray source and the detector are rotated around your body, slowly moving from your head to your feet until every inch of your body has been scanned from all directions. All of the differences in X-ray intensity detected by each little scintillator piece are gathered by the computer and reconstructed into a three-dimensional map of your body. The doctor can then instruct the computer to show the reconstructed image of

Figure 8-10. *Transparent ceramic scintillator material and other scintillator parts* (Samples supplied by GE Medical Systems, Milwaukee, WI. Photo by D. Richerson.)

any cross section of your body on a television screen, like the pictures shown in Figure 8-8. Although CAT scans are expensive, they give doctors an enormous amount of information that previously was available only through exploratory surgery.

Ultrasonic Imaging

Ceramics are also the heart of the medical *ultrasound* scans that doctors use to show a mother the baby developing in her womb. The "ultrasound" consists of ultrasonic waves generated by a ceramic piezoelectric transducer similar to those discussed in Chapter 7 for nondestructive testing. Electricity coming into the transducer causes it to vibrate, producing sound waves that travel into the woman's body. Some of the sound waves are reflected by any new surface they encounter (such as a bone or organ); others continue through the body. The reflected waves are detected by the ceramic piezoelectric transducer and converted to electrical signals that can be changed into an image and shown on a screen.

Endoscopy

Forty years ago, the only way a doctor could diagnose an internal medical problem was through major exploratory surgery. Now, doctors can use not only CT and ultrasonic imaging but also *endoscopy*. Endoscopes can be used to examine the inside of the colon (your large intestine) and the female reproductive system, especially to search for cancer. They're also important for directly locating, and even surgically correcting, knee damage, problems in the abdominal cavity (such as gall bladder, appendix, or an unknown source of internal bleeding), and even the insides of your blood vessels and heart valves.

A diagnostic endoscope is essentially a flexible tube that the doctor can insert into your body and look through. Since the inside of your body is dark, the endoscope must have a light source, so part of the tube is filled with glass optical fibers attached to a bright light. Other glass optical fibers are attached on one end to an eyepiece that the doctor looks through, and on the other end to a glass lens at the end of the endoscope. Some endoscopes have a built-in mechanical manipulator like a miniature robotic hand, surgical tools like a scalpel, and even a channel for flowing water or air, so that the doctor has an unobstructed view through the lens. Because an endoscope can enter your body through normal openings such as your mouth or rectum or through a small incision, the endoscope is a big help to your doctor in accurately diagnosing internal ailments without the risk and long recovery time of major invasive surgery.

CERAMICS IN MEDICAL TREATMENT AND THERAPY

Ceramics are important in just about every aspect of medical treatment, from surgery to physical therapy to the manufacture of pharmaceuticals. They're even important in cancer treatment, dialysis, and heart stabilization with pacemakers.

Laser Surgery

As we discussed in Chapter 4, ceramics are an integral part of most lasers. The list of laser uses in medicine is growing every year. Lasers are already well-established for eye surgeries, and they also have been approved recently for dental work, as a painless alternative to the dental drill. Another area gaining in popularity is skin repair. The outer layers of your skin that have been damaged by years of exposure to the sun actually can be vaporized by the laser, allowing fresh tissue and skin to grow back in their place.

Traditional Surgery

Ceramics are used in traditional surgery for everything from scalpels to endoscopes to high-intensity lighting for operating rooms. Scalpels made of ceramics are very sharp and cut smoothly and evenly. Doctors have even attached a scalpel to a piezoelectric transducer to produce an

USES OF CERAMICS IN MEDICAL TREATMENT AND THERAPY
Surgical lasers
Tools for traditional surgery
Microspheres for cancer treatment
Dialysis equipment
Devices for unblocking blood vessels
Physical therapy equipment
Pacemakers
Respirators
Pill presses
Pharmaceutical metering devices
Microbeads for live-cell encapsulation

Figure 8-11. *Fixture for high-intensity lighting in operating rooms.* *(Sample supplied by Coors Ceramics, Golden, CO. Photo by D. Richerson.)*

ultrasonic scalpel that gives unprecedented control and precision in delicate operations.

Endoscopic surgery has really grown in importance in recent years. Many of you have probably had endoscopic knee surgery to repair cartilage or ligament damage from jogging, football, or skiing. The doctor makes a small incision on one side of your knee, through which he or she can insert the endoscope to inspect the damage and guide the surgery. The surgical tools are inserted through one or more separate incisions. Another breakthrough has been endoscopic gall-bladder surgery. Every year, surgeons seem to come up with a new endoscopic procedure that reduces surgery time, avoids having to completely cut open the patient, reduces the risk of the operation, and greatly decreases recovery time and discomfort.

Removing potentially cancerous polyps from the colon is another important endoscopic procedure. Polyps are small outgrowths that can grow on the inside of your colon. They're often a first step to cancer and so should be detected as soon as possible and removed. The doctor looks inside your colon with a fiber-optic endoscope to locate and examine polyps. The endoscope has a tiny ceramic tool (an electrical insulator) at the end, tipped with a thin spiral of metal paint (see

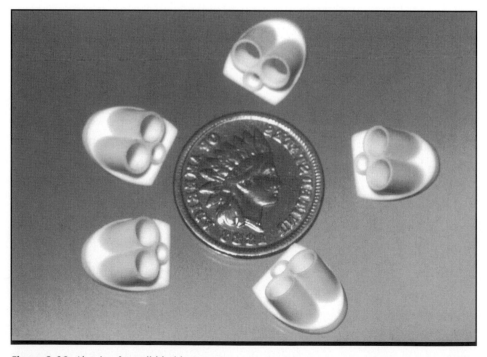

Figure 8-12. *Alumina for gall-bladder surgery.* *(Samples fabricated by Lone Peak Engineering, Draper, UT. Photo by D. Richerson.)*

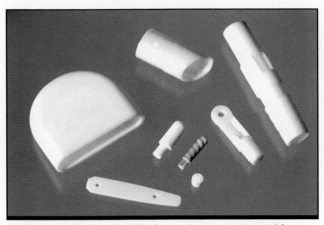

Figure 8-13. *Variety of medical ceramic parts, some used for surgery.* (Samples supplied by Coors Ceramics, Golden, CO. Photo by D. Richerson.)

Figure 8-13) specially designed to get hot when electricity is passed through it. A metal wire to carry electricity goes through the endoscope and the ceramic tool and is connected to the metal spiral. When the doctor sees a polyp through the endoscope, he or she touches the base of the polyp with the metal spiral and turns on an electrical switch. The spiral heats up and burns off the polyp, cauterizing (closing off blood vessels to prevent bleeding) the wound at the same time.

Radioactive Glass Microspheres for Cancer Treatment

Radiation is effective for destroying cancer cells, but unfortunately it also kills normal cells. Doctors are faced continually with the challenge of delivering radiation to our bodies to kill the cancer cells while minimizing damage to surrounding healthy cells. Traditionally, doctors have destroyed cancer cells by irradiating them from the outside of a patient's body with a beam of high-intensity gamma rays, but that method can't avoid damaging the healthy tissue that the beam must go through to get to the cancerous tumor. Dr. Delbert Day and his co-workers at the University of Missouri-Rolla have developed an alternative approach that shows great promise. They use tiny radioactive glass spheres instead of gamma rays, in an innovative approach to the battle against cancer. This treatment has been so effective that some liver-cancer patients who normally would survive less than six months are now still alive and healthy after six years.

Day and his associates discovered how to make tiny microspheres of glass, about one-third the diameter of a human hair, that can be made radioactive in a special way. Rather than emitting beams of radiation that penetrate long distances, the glass spheres emit radiation that can be controlled to reach out only short distances, such as one to ten millimeters (less than one-tenth of an inch to less than one-half inch). This feat is accomplished by controlling the chemical composition of the glass spheres. In addition to controlling the distance that the radiation can reach out around each sphere, the University of Missouri-Rolla team also has been able to control the length of time that the glass microspheres stay radioactive.

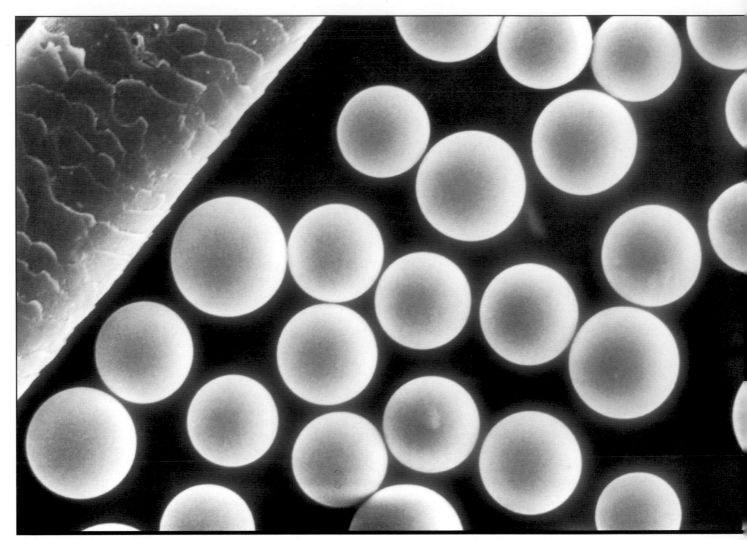

Figure 8-14. *Radioactive glass microspheres for cancer treatment. The size of the microspheres is compared with that of a human hair.*
(Scanning electron microscope photo courtesy of Delbert Day, University of Missouri, Rolla.)

Interesting, you say, but how can these radioactive microspheres be used? Suppose you're diagnosed with liver cancer. The doctors run a CAT scan to map out the size of tumors in your liver. They calculate how far the radiation from microspheres trapped in the tumors would have to reach out to attack all of the cancerous tissue with minimum attack on surrounding tissue. They then place an order for these "designer" microspheres. When the microspheres are delivered to the doctor, they're placed in a fluid and injected into your femoral artery, the blood vessel that leads directly to your liver. Tumors are blood hogs; they take a large percentage of the blood flow to an organ. The radioactive microspheres (about 5 to 10 million) enter your liver and are filtered out of the blood by the tiny blood vessels (capillaries) in the tumor, where they start to do their work.

Glass microspheres give doctors important advantages in the fight against cancer. By delivering the radiation directly to the inside of your tumor, they can apply a radiation dose five to seven times higher than is possible with externally applied gamma radiation and without side effects or discomfort. Usually, you need to undergo only one treatment lasting two to four hours, compared to chemotherapy that lasts for weeks and has severe side effects. The glass microspheres kill the tumor and completely lose their radioactivity in three to four weeks, and they don't appear to have any adverse side effects.

Approved in Canada and Hong Kong, glass microsphere radiation treatment for liver cancer is expected to be approved soon in the United States, Europe, and China. Radioactive glass microspheres are also being explored for treating kidney cancer and brain tumors. Similar radioactive glass microspheres, but that are biodegradable and gradually dissolve, are being evaluated as a treatment for rheumatoid arthritis.

Dialysis

Dialysis is a medical procedure that keeps you alive if your kidneys no longer work. Your body is like a complex chemicals factory. Oxygen from the air you breathe and the food you eat is processed by your lungs, stomach, and other organs and carried by blood cells to every cell in your body. Tiny chemical reactions take place in your cells to fuel your muscles and other systems. These reactions give off chemical by-products that are toxic to your body and must be removed. Many of these toxins are carried by your blood to your kidneys, where they are filtered out and discharged in urine.

If your kidneys are damaged by accident or disease and stop working, you will die. Dialysis is a medical procedure that can keep you alive until either your kidneys heal or you receive a kidney transplant. During dialysis, blood from your body is circulated

Figure 8-15. *Piezoelectric ceramic parts for medical applications. Samples at lower left and right, with wires connected, are for kidney dialysis bubble detectors. Dark samples left of center are for physical therapy.* (Photo courtesy of Edo Corporation Piezoelectric Products, Salt Lake City, UT.)

through the dialysis machine to be purified and then returned back into your body. A major danger during dialysis is that air bubbles will get into your blood and trigger a heart attack or stroke. To prevent such crises, piezoelectric ceramic transducers are set up on each side of the tubing carrying the blood, one to continually send sound waves through the tubing and blood, and the other to continuously monitor those sound waves. Any nonuniformity in the blood flow, such as a bubble, disrupts the steady stream of sound waves between the ceramic piezoelectric transducers and sets off an alarm.

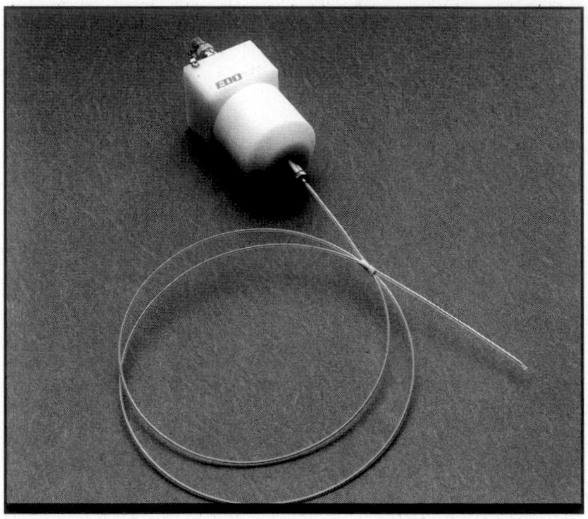

Figure 8-16. *Miniature piezoelectric "jackhammer" for unblocking blood vessels.* (Photo courtesy of Edo Corporation Piezoelectric Products, Salt Lake City, UT.)

Unblocking Blood Vessels

Buildup of fatty deposits in blood vessels restricts blood flow and is a major cause of cardiovascular (heart, lung, and circulatory system) distress. Studies have demonstrated that a piezoelectric device that's essentially a "microjackhammer" can be inserted into a clogged blood vessel to successfully erode away fatty coatings and open up the blood vessel. This procedure has the potential to be cheaper, simpler, and safer than present surgical techniques.

Physical Therapy

Ultrasound medical treatment was introduced in Germany in the late 1930s and in the United States in the late 1940s. Ultrasonic vibrations cause a slight increase in the temperature deep inside your muscles and in the ligaments that connect these muscles to your bones. The heat promotes healing, increases flexibility, and decreases pain.

The most common ultrasonic therapy is called *heat and stretch*. When you're injured or have surgery such as knee repair, your muscles and joints become very stiff and painful to move. Your therapist will force you to bend and stretch farther and farther, until you get back your original range of motion. This stretching is normally a slow and painful experience, but it can be speeded up and made less painful with ultrasound therapy. The therapist places a small, hand-held instrument with a built-in, flat ceramic piezoelectric disk roughly one to one and a half inches (2 to 4 centimeters) in diameter against the injured region. The ultrasonic vibrations from the piezoelectric ceramic deep-heat the tissue, all the way to the muscle-bone connections, to a temperature of 102° to 117°F (39 to 47°C). This heating makes tissues such as collagen more pliable and increases your range of motion during stretching exercises. The more quickly you can regain your range of motion, the more quickly you can return to normal activities.

Similar levels of ultrasound heating have been shown to reduce pain from muscle spasms and to ease the symptoms of bursitis and tendinitis. Ultrasound therapy also has been helpful in treating lower-back injuries and pain, in enabling calcium deposits in soft tissue to be reabsorbed into the body, and in breaking down scar tissue.

Ultrasound has been successful for a variety of other treatments. Driving anti-inflammatory drugs (such as cortisol, salicylates, and dexamethasone) through the skin to the inflamed underlying tissue has been one success. For example, cortisol was driven to a depth of 2 to 2.4 inches (5 to 6 centimeters) in skeletal muscle. Some success has been demonstrated in destroying cancer tissue with high intensity ultrasound. Conversely, low-intensity ultrasound, resulting in a temperature rise of only about 0.9° to 2.4°F (0.5 to 1.35°C) has increased blood flow and aided in healing.

Ceramics in Pacemakers

Your heart is an amazing organ. It beats at a constant speed day after day, year after year. Sometimes, though, especially as you get older, your heartbeat can become dangerously fast. Doctors have learned that mild electrical shocks to the heart can slow it down to a normal speed. A *pacemaker* is a device that can be surgically implanted into your chest to continually release electrical pulses to control your heart rate. Ceramic materials are important parts of every pacemaker, working as electrical insulators and capacitors and also as glass seals to keep body fluids from getting inside the pacemaker. Even the outer case (enclosure) of some pacemakers is now made of a ceramic (such as alumina), because ceramics are more compatible with our bodies than other materials.

Ceramics in Respirators

The field of medicine has seen amazing advances in recent years toward keeping alive people who are not able to breathe on their own because of severe injuries, disease, premature birth, or complex surgical procedures. A key piece of equipment in this battle is the *respirator*. It breathes for the person by forcing small "gulps" of air or oxygen into the lungs. To do this, the respirator machine must have a little valve that slides back and forth with each breath of air. This valve must move smoothly and without the use of a lubricant (such as oil), must have such a small clearance between the moving parts that air doesn't leak out, and must slide back and forth millions of times without sticking or wearing. Ceramics such as alumina can meet these challenges.

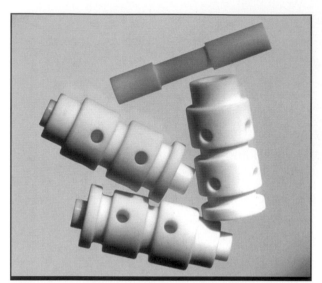

Figure 8-17. *Alumina ceramic oxygen-valve cylinder and piston for metering oxygen flow in a respirator.* (Sample fabricated by Coors Ceramics, Golden, CO. Photo by D. Richerson.)

Pharmaceutical Uses for Ceramics

Have you ever wondered how pills are made? They're made from powders that are pressed at high pressure in metal dies. The die is shaped like a toilet-paper roll but much smaller. A measured amount of powder of the desired pill composition, mixed with a little gluelike *binder* is poured into the hole in the center of the die. Metal rods are pushed into the hole from both ends, under enough pressure to make the powder and binder stick together to form a solid pill. The finished pill then is pushed out of the die by the bottom rod, and the cycle is repeated. Pill presses work at lightning speed to make many pills each minute. The inside of the hole in the die often is lined with a hard ceramic material to increase the life of the die and prevent metal particles from scraping off and discoloring the pills. Other parts of the press, such as the device that controls the amount of powder metered into the die for each pill, also are sometimes made of ceramics.

Live-Cell Encapsulation

New important applications of ceramics in medicine are being reported each year. One exciting possibility for the future is *live-cell encapsulation,* to implant genetically engineered cells into the body as biotech "factories" to produce chemicals such as insulin. The foreign cells must be protected from the body's immune system, yet allowed to exchange nutrients and waste products and to secrete their manufactured chemicals. Edward Pope of Solgene Therapeutics, in Westlake Village, CA, has demonstrated that healthy pancreatic islets (cells that produce insulin) encapsulated in sol-gel silica ceramic microbeads and implanted in diabetic mice can successfully produce the insulin needs of the mice. Effort is underway to see if this works in humans.

Figure 8-18. *Pharmaceutical metering device for measuring pill powders into tooling for pressing into pills.* (Sample fabricated by Coors Ceramics, Golden, CO. Photo by D. Richerson.)

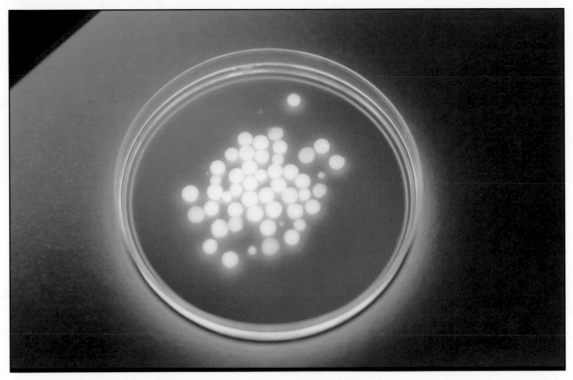

Figure 8-19. Sol-gel ceramic microbeads for live-cell encapsulation. *(Photo courtesy of Solgene Therapeutics, Westlake Village, CA.)*

OVERVIEW: MEDICAL MIRACLES YET TO COME

Ceramics have had a profound impact on medicine just during the present generation. Such uses for ceramics as in the glass-microsphere treatment of cancer, laser surgery, endoscopic diagnosis and surgery, and high-resolution CT scans not only seem magical but also are truly miraculous. You can look forward to many more ceramic medical breakthroughs during the next generation. Radioactive glass microspheres for liver cancer treatment will be approved in the United States and will lead to similar treatment of other cancers. Joint implants other than hips will be refined and become as commonplace as hip implants. Genetic engineering will continue to grow and will use ceramics as well as plastics for the live-cell encapsulation of miniature biotech factories in our bodies. And someday, ceramic sensors and advanced electronics probably will continually monitor our health and provide feedback, by way of a miniature telemetry transmitter to a hospital computer that can alert a doctor if dangerous symptoms arise.

The next chapter follows the same theme as this chapter: exploring the uses of ceramic materials in a particular field. The automotive field has been selected, which seems to be nearly as important to many people as their health.

Ceramics and the Modern Automobile

CHAPTER NINE

Most of you ride in cars every day and don't realize the number of ways ceramic materials are used in them. As you learned in Chapter 2, ceramic spark plug insulators played a critical role in the development of early cars, but now ceramics are used in at least a hundred ways, from the electronic engine control-system to the motors that run the power seats to the sensors that tell how much gas you have in the gas tank. They are the heart of your catalytic converter pollution-control system, which has saved the world from 1.5 billion tons of air pollution since 1976. They are the source of light for the instrument panel, the key material in the sensors that tell you whether your car is healthy or not, and an important part of your car's sound system. Modern automobiles wouldn't be possible without ceramic materials. This chapter will introduce you to uses for ceramics under the hood, in the passenger compartment, and in other areas of your car and also describe some of the ways that ceramics are used to manufacture cars.

> **PRODUCTS AND USES**
>
> Spark plugs
> Oxygen sensors
> Catalytic converters
> Electronic engine-control systems
> Water-pump seals
> Air-conditioner compressor seals
> Turbocharger rotors
> Sensors
> Thermal insulation
> Fiber-reinforced composites
> Diesel glow plugs
> Diesel injector timing plungers
> Diesel cam follower rollers

CERAMICS UNDER THE HOOD

When asked where ceramics are used in the engine or other places under the hood of their car, most people say in spark plugs but can't think of even one other use. Maybe the reason is that the white alumina ceramic insulator in a spark plug is easily visible, whereas all of the other ceramic parts are hidden inside the engine, the water pump, the air-conditioner compressor, the pollution-control system, and other places. Let's take a closer look to see where and why ceramics are presently being used under the hood of gasoline- and diesel-powered motor vehicles. Then we'll look

at some new ways that automotive engineers are trying to use ceramics to decrease the amount of fuel that cars use, to further cut down on the air pollution given off, and to make the cars run longer with less maintenance.

Oxygen Sensors and Catalytic Converter

By 1970, high-quality spark plugs had engines purring, and the world's roads and highways were buzzing with automobiles, trucks, and buses, especially in and around major cities. Motor vehicles also were producing 200 million tons of pollution per year in the United States alone. Americans came to realize that automobile pollution was turning our major cities into respiratory death traps. In response to this severe health hazard, the U.S. Congress passed The Clean Air Act in 1970, with the intent of reducing pollution emissions from automobiles and other vehicles by 90 percent. The development of oxygen sensors and catalytic converters, both completely dependent on ceramics, enabled car manufacturers to meet that goal in less than 10 years.

All automobiles with standard internal-combustion engines in the United States, Europe, and Japan are now required by law to have oxygen sensors. Mounted in the exhaust manifold, an *oxygen sensor* continually monitors the oxygen content of the combustion gases that leave a car's engine. This residual oxygen content indicates how efficiently the fuel is burning and the engine is running. The oxygen sensor not only detects the oxygen content of the hot gases leaving the engine but also sends an electrical signal to the electronic engine-control system. If the signal indicates that the engine is not operating at peak efficiency, the electronic engine-control system then readjusts the amount of air that's mixed in with the fuel, so that the engine again works under optimum conditions. Fortunately, the conditions that produce minimum pollution also result in peak gas mileage. The oxygen sensor gives, on average, about four extra miles per gallon of gasoline, probably enough for five to ten trips to the grocery store on each tank of gas.

GASOLINE SAVER

Ceramic oxygen sensors actually increase gas mileage by about four miles per gallon.

The key element in an oxygen sensor is a special type of electrically conductive zirconium oxide ceramic. Rather than conducting electrons the way a metal does, this special zirconia conducts oxygen *ions*. An oxygen ion is an oxygen atom that has picked up two extra electrons. Because it now has more electrons than protons, it has a negative electrical charge and can be a carrier of electricity. An oxygen ion, however, is thousands of times larger than an electron and can't travel through metal wire; it can only travel through special ceramics, such as the zirconia.

The special zirconia used for an oxygen sensor is fabricated into a hollow thimble shape, as shown in Figure 9-1, and coated on the inside and outside with a thin, but porous, layer of an electrically conductive metal that's chemically stable in the hot exhaust gases of your car. These metal layers (*electrodes*) are connected to an electrical device that can measure the presence and amount of an electrical voltage. Now, here's the secret of the oxygen-sensor. The inside of the oxygen-sensor thimble is exposed to fresh air, which has a known content of oxygen. The outer surface is exposed to the hot exhaust gases from the engine, which have a different content of oxygen than fresh air. Because of this difference in oxygen content, oxygen ions try to travel through the zirconia from the higher-oxygen side to the lower-oxygen side, by a sort of osmosis. This results in a measurable electrical voltage. The larger the difference in oxygen content between the fresh air and the exhaust, the larger the voltage. Through years of study, engineers have learned the exact amount of oxygen in the exhaust that gives minimum pollution and gas consumption, so that they know the optimum voltage. Whenever this voltage is given off by the oxygen sensor, the electronic control system knows that the engine is running just fine. If the voltage changes, the electronic control system is programmed to cause the fuel-air mixture to change until the voltage comes back into an acceptable range.

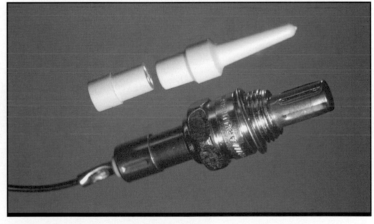

Figure 9-1. *Oxygen-sensor ceramic pieces and assembly.* (*Samples from AlliedSignal Filters and Spark Plugs, Fostoria, OH. Photo by D. Richerson.*)

The second key to motor-vehicle pollution control is the *catalytic converter*. A catalytic converter consists of a ceramic honeycomb substrate coated with a thin layer of a catalyst and mounted in a metal chamber that funnels all of the hot engine exhaust gases through the coated honeycomb. A cutaway cross section of a catalytic converter is shown in Figure 9-2. A *catalyst* is a material that encourages chemical reactions to occur but is not involved in the reaction and is not consumed. The catalyst is specially designed to cause the pollutants carbon monoxide, nitrogen oxide, and unburned hydrocarbon compounds (soot) to be chemically converted to harmless water, nitrogen, and carbon dioxide.

Figure 9-2. *Cutaway view of a catalytic converter.* (Photo courtesy of Corning, Inc., Corning, NY.)

The ceramic honeycomb substrate in a catalytic converter must have thin walls, contain many cells, and also resist the frequent and drastic temperature changes that occur inside an automobile engine. The thin walls and numerous cells are necessary to provide a large surface area, to assure that a high percentage of the pollutants are decomposed. Modern ceramic substrates have about 350 to 400 cells per square inch (54 to 62 cells per square centimeter) and a wall thickness of about 0.0055 inch (0.14 millimeter). They are a marvel of manufacturing technology. A complete substrate is shown in Figure 9-3, and a close-up showing the small size of the openings and the thinness of the walls is shown in Figure 9-4.

Normal ceramics can't withstand the thermal shock imposed by the rapid thermal cycles of the catalytic converter, especially when you start your car on a cold winter morning. To solve this problem, Corning, Inc. developed a material,

Figure 9-3. *Ceramic honeycomb substrate for a catalytic converter.* (Sample from Corning, Inc., Corning, NY. Photo by D. Richerson.)

called *cordierite,* with near-zero thermal expansion. Since 1974, over 350 million cordierite ceramic-honeycomb catalytic converters have been installed in cars, trucks, and buses and have reduced pollution by 1.5 billion tons.

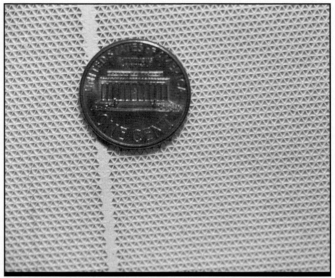

Even though the traffic in our major cities has increased enormously in the past 30 years, the overall pollution level is much lower than in the early 1970s, before the oxygen sensor and catalytic converter were installed in our cars. New developments in catalytic converters, oxygen sensors, and high-temperature ceramic particle filters (to filter out the smoke and soot from diesel engines) should further decrease pollution emissions from motor vehicles. At the rate technology is progressing, these com-

***Figure 9-4.** Close-up view of cordierite ceramic honeycomb. (Photo by D. Richerson.)*

bined ceramic systems might someday be so efficient that they could actually make the emissions coming out of our cars cleaner than the air entering our engines. What a dramatic change that would be; we could use our cars to cleanse our air!

Electronic Engine-Control System

The brain that coordinates and manages engine control and all of the electrical systems in a motor vehicle, including the spark plugs and the oxygen sensor, is the *hybrid ceramic electronics system.* This system consists of half a dozen modules made up of complex silicon chip integrated circuits, plus other ceramic and semiconductor devices mounted on the surface of

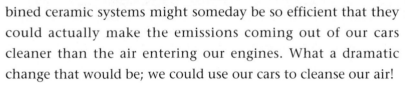

thin sheets of aluminum oxide. As shown in Figures 9-5 and 9-6, the alumina has a pattern of metal lines printed on the surface to interconnect the silicon chips and other electrical components. Over 100 million modules are produced each year in the United States.

Central to the engine-control system is the master control module, which is essentially a computer only about 3 inches (7.6 centimeters) wide and 4 inches (10.2 centimeters) long. This module is mounted between the fire wall (which isolates passengers from the engine heat and provides protection in case of fire in the engine compartment) and the dashboard in cars, and right inside the engine compartment for marine engines, so it must be rugged and reliable. Present master control modules are designed for operation between -104° and +176°F (-40° and +80°C). Future automotive

SAVING OUR AIR

Automotive catalytic converters have reduced pollution by 1.5 billion tons since 1974.

Experimental systems have demonstrated zero pollutant emissions.

Motor-vehicle pollution control will be a $38 billion business by the year 2000.

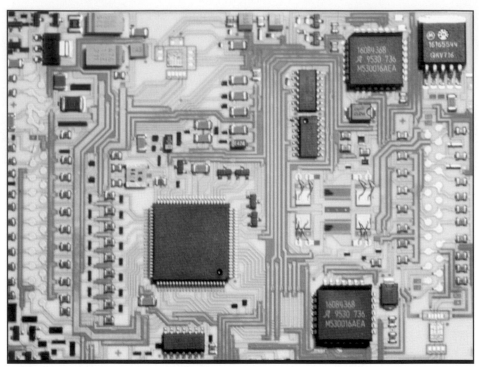

Figure 9-5. *Electronic master control module, showing silicon chips, capacitors, resistors, and metal wiring all integrated onto a thin sheet of alumina ceramic.* (Sample from Delco Electronics, Kokomo, IN. Photo by D. Richerson.)

modules will be mounted in the engine compartment and are being designed to endure temperatures ranging from −122° to +302°F (−50° to +150°C).

Other electronics modules, which are smaller than the master control module, fulfill various roles in your car's engine. The voltage-regulator module is sealed in the generator and controls the charging of the battery. The electronic spark-control module discerns damaging engine knock from other engine noise and sends a signal to the master control module to make adjustments to the engine timing. Linked to the oxygen sensor and a heated-wire airflow-measuring device, an air-meter module provides signals to the master control module to adjust the air–fuel mixture. A manifold air-pressure module calculates the mass of air entering the engine, and a direct-ignition module controls spark and timing. All of these modules work together with the oxygen sensor to keep your car running smoothly, help reduce air pollution, and optimize gas mileage.

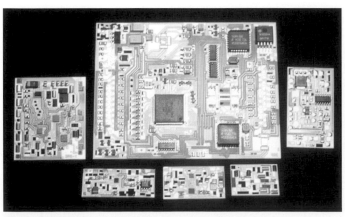

Figure 9-6. *Master control with other electronics modules.* (Samples from Delco Electronics, Kokomo, IN. Photo by D. Richerson.)

As you can see, the hybrid ceramic electronics system in your car is quite remarkable. Compact and amazingly durable, the modules survive bumps and vibration, weather extremes, and even failures of other components such as the battery, generator, and voltage regulator. The system usually performs invisibly and trouble-free for the life of your automobile.

Water-Pump and Air-Conditioner Compressor Seals

Most of our car engines run hot enough that they must be continuously cooled by water pumped through channels built into the engine. The water heats up as it flows through the engine and is cooled in the radiator. The pump that circulates water back and forth between the radiator and the engine has a ceramic seal that keeps the pump from leaking during each pumping cycle. The ceramic seal is a relatively simple ring with a very smooth surface that can seal and slide simultaneously. Examples of water-pump seals made of silicon carbide are shown in Figure 9-7. Seals were made for many years out of alumina, but many are now made of silicon carbide, because it's harder than alumina and lasts longer. Some seals also have been made of silicon nitride, which is tougher than either alumina or silicon carbide. Similar ceramic seals also are used in the compressor of the car's air-conditioning system.

Figure 9-7. Water-pump seals fabricated from silicon carbide, for water pumps in cars and trucks. Greater than 20 million produced each year. (Samples from Saint-Gobain/Carborundum Structural Ceramics, Buffalo, NY. Photo by D. Richerson.)

Turbocharger Rotors

A spectacular application of advanced ceramics, involving a complex shape (shown in Figure 9-8) and high stresses, has been for turbocharger rotors. These rotors were introduced in 1988, in Japan, as an outgrowth of the development of silicon nitride gas turbine engine components. Located in the engine exhaust, a turbocharger extracts some of the energy from the hot engine gases, which normally would be wasted, and uses that energy to boost the power and performance of the car's engine.

Turbochargers with metal rotors had been standard equipment on trucks for many years and were introduced to some cars in the 1970s and 1980s. The sales appeal of a turbocharger was that the extra energy boost it supplied would give a four-cylinder engine as much zip as a six-cylinder engine. Drivers weren't as impressed as the auto manufacturers expected, though. The metal rotor was so heavy that it caused a lag, or hesitation,

Figure 9-8. *Silicon nitride turbocharger rotor that spins in an automobile up to 140,000 rpm and extracts extra energy from the hot engine exhaust gases.* (Photo courtesy of Kyocera.)

when the driver stepped on the accelerator, rather than allowing the car to immediately leap forward with maximum power. At about one-third the weight of a metal rotor, a silicon nitride ceramic rotor could accelerate to speed much more quickly and thus didn't irritate the driver with hesitation or lag. Turbochargers with ceramic rotors became very popular in Japan, where approximately 30,000 ceramic turbocharger rotors were produced per month starting in 1988.

Ceramic Sensors

Many of the ceramics in your car's engine compartment are used as sensors, especially those ceramic semiconductor compositions that change their electrical properties with changes in temperature or voltage. For example, a *thermistor* changes electrical resistance at different temperatures and can be used as a temperature sensor, temperature compensator, switch, or heater. A *varistor* has high electrical resistance at low voltage and low resistance at high voltage and can be used to protect electrical devices from occasional high-voltage surges. Such ceramic materials are used in the blower motors, the intake air sensor, the cooling-water temperature sensor, the oil-temperature sensor, and the knock sensor.

Ceramic Fibers and Composites

Also important in the engine compartment are ceramic fibers, both as loose bundles used for thermal insulation (like the fiberglass insulation in your house) and as reinforcement in polymer-matrix composites (like those we discussed in Chapter 5). Most of the fibrous thermal insulation is in the fire wall between the engine compartment and the passenger compartment. The composites are made up of glass fibers that have been chopped into short lengths and embedded in polymers. One application is for the frame that supports the radiator in cars such as the Ford Taurus and Mercury Sable. Another is in the air-intake manifold, the complex ductwork that carries the air from the radiator to each engine cylinder. This complex-shaped manifold is expensive and difficult to make from a metal but relatively easy and inexpensive to mold from glass-fiber-reinforced plastic.

Ceramics in Diesel Engines

Production diesel engines also use ceramics. Two such applications are in the fuel heater (prechamber) and the glow plug. If you've ever driven a diesel car, you know that you have to wait 15 to 30 seconds after turning on the switch before you can push the starter. This delay is the time it takes to preheat the fuel. Diesel fuel is thicker and heavier than gasoline and doesn't vaporize as easily, so is preheated with a ceramic heater before it can be ignited with a ceramic glow plug. Both of these devices are fabricated from silicon nitride.

The metering of diesel fuel is also a challenge and requires a precision fuel-injection system. The key component of the fuel-injection system is the *timing plunger*. Timing plungers used to be made of metal and were a major maintenance problem because they would wear and no longer meter the fuel properly. Metal timing plungers were replaced with silicon nitride in 1989. New designs introduced in 1995 were made with transformation-toughened zirconia. Approximately 3 million ceramic timing plungers have been installed in truck engines and appear to have solved the wear and maintenance problems of the prior metal parts.

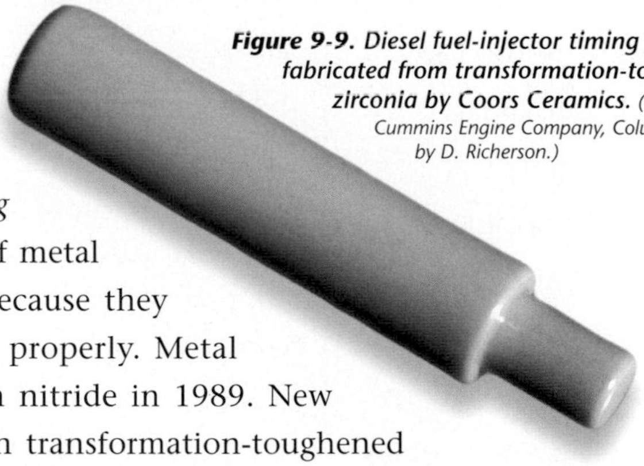

Figure 9-9. *Diesel fuel-injector timing plunger fabricated from transformation-toughened zirconia by Coors Ceramics. (Sample from Cummins Engine Company, Columbus, IN. Photo by D. Richerson.)*

Another production diesel use for ceramics is as cam follower rollers fabricated from silicon nitride. The up-and-down motion of the pistons in the engine must be transformed into the rotating motion of the drive shaft. Previously, this transformation was achieved by mechanical levers and a sliding motion. The sliding interfaces, however, wore too quickly

Figure 9-10. *Diesel-engine silicon nitride cam follower rollers. (From Kyocera Industrial Ceramics Corporation, Vancouver, WA. Photo by D. Richerson.)*

and also decreased the power from the engine, because of friction. A ceramic roller design at Detroit Diesel replaced the sliding metal design. The silicon nitride rollers have reduced engine wear and friction, resulting in longer life and improved engine performance.

Experimental Ceramic Engine Components

The U.S. government and, especially, the state of California have continued to pass legislation requiring further reductions in pollution and increases in gas mileage, and automotive engineers are always trying to improve the engines and systems to make their companies' cars more competitive. As we've discussed, ceramics have played a major role in helping meet pollution reduction, but they also offer automobile manufacturers other important advantages that can increase engine performance and gas mileage: higher temperature capability, better wear resistance, greater ability to work with little or no lubrication, and light weight. Researchers all over the world are busily building automotive parts out of ceramics and testing them in the laboratory and in cars. The following paragraphs tell about some of these efforts, which will surely lead to new ceramics uses under the hoods of motor vehicles.

EXPERIMENTAL CERAMIC ENGINE COMPONENTS

Valves, guides, and seats

Pistons or piston caps

Cylinder liners

Piston pins

Exhaust-port liners

Manifold insulation

Wear-resistant coatings

Thermal-barrier coatings

Alternative turbine engines

Piston bases for Stirling engines

Rocker arms

Ceramic Valves. A valve is a piece of material that moves in and out of an opening to either block or allow flow through the opening. Valves are used in gasoline and diesel engines to let the air–fuel mixture into the cylinder that houses the piston and also to let out the hot exhaust gases after the air–fuel mixture has been ignited by a spark from the spark plug. The valves, which are presently made of metal, are exposed to the full force and temperature of the exploding gas–fuel mixture. Replacing the metal valves with a higher-temperature and lower-weight material would increase valve life and increase engine efficiency.

Ten years ago, automotive manufacturers didn't believe that ceramics could survive as an engine valve, but they were wrong. Silicon nitride, the same high-strength ceramic that works so well for bearings and cutting tools, also works beautifully for engine valves. Silicon nitride valves have been run in high-performance race car engines, where materials are pushed to their limits. They have also been run in passenger cars by several different automotive companies. The most aggressive automotive company has probably been Daimler Benz. They've successfully tested silicon nitride valves in engines in the laboratory and on the road for hundreds of thousands of miles and have proved that these ceramics meet all of the reliability requirements. The only thing keeping them from putting ceramic valves into production is cost, but Daimler Benz is also getting very close there. Don't be surprised to see cars with silicon nitride valves available in your dealer's showroom within the next few years.

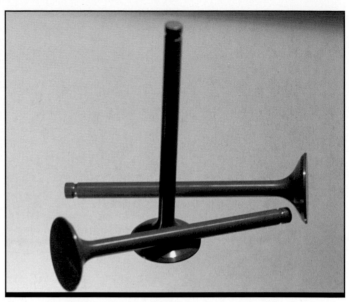

Figure 9-11. *Silicon nitride experimental automobile valves.* (Samples from Saint-Gobain/Norton Advanced Ceramics, Northboro, MA. Photo by D. Richerson.)

Ceramic-reinforced Pistons. Pistons are another important candidate for ceramics, but most companies aren't pursuing a solid ceramic piston. Instead, they're exploring the use of ceramic particles, whiskers, or fibers as reinforcement in aluminum alloys. The ceramic reinforcement increases the strength and temperature resistance of aluminum, enabling manufacturers to substitute lightweight reinforced aluminum for heavy iron alloys and substantially reduce the weight of the car's pistons. The improvements in strength and temperature capability can be quite remarkable. For example, replacing only one-third of the aluminum with silicon carbide ceramic whiskers increased the strength of one aluminum alloy from about 10,000 psi to over 90,000 psi, and the alloy retained the increased strength to over 750°F (400°C).

Other Ceramic-reinforced Metal Parts. Whisker reinforcement can be costly, so researchers have focused considerable attention on adding

amazing facts

A mere 30 percent addition of silicon carbide ceramic fibers to one aluminum alloy increases the strength of the alloy from 10,000 to 90,000 psi.

inexpensive particles of aluminum oxide or silicon carbide, rather than whiskers, to aluminum. Although the properties of parts reinforced this way haven't been as high-grade as when whiskers have been used, the parts nevertheless seem adequate for many automotive components. For example, aluminum reinforced with ceramic particles was made into car engine connecting rods and proved to have higher strength and more resis-

Figure 9-12. *Ceramic-metal composite experimental automotive parts.* *(Photo courtesy of Lanxide Automotive Products, Newark, DE.)*

tance to the stresses imposed by the engine than the metal alloy currently used.

Another aluminum alloy with alumina particles as reinforcement looks promising for long drive shafts. Long drive shafts in trucks and some planned cars have a rotational stability problem at high speed—that is, they start to vibrate like a rubber band that's been plucked. Adding 20 percent alumina particles makes the drive shaft about 36 percent stiffer, enough of an increase in stiffness to keep the drive shaft from vibrating at high speed.

Ceramic-reinforced metals have great potential, but cost is a problem. In most cases, automotive companies won't use a new material until they can get the cost down to what they are paying for the current material. Some innovative fabrication processes have been developed to decrease the cost of ceramic-reinforced metals. One process involves heating a block of the composite to a temperature at which the block can be *forged* (deformed under pressure and heat into a special tool that has the shape of the desired part). Another process loosely bonds together ceramic particles into a porous *preform* of the desired shape and then lets molten metal soak into the preform and fill the pores. In this process, the ratio of ceramic to metal can be varied anywhere from 40 parts ceramic/60 parts metal to about 80 parts ceramic/20 parts metal, so that a whole family of materials with different properties is possible. A third process, which only works for up to about 30 to 40 percent ceramic particles, involves heating the metal–ceramic mixture to above the melting temperature of the metal and pouring the thick-as-molasses mixture into a shaped ceramic mold to cool and become solid.

Ceramics in Experimental Diesel Engines. Diesel engines run hotter than gasoline engines. This allows them to be more efficient but also results in a lot of energy tied up as heat. In conventional diesel engines, this heat and the energy it contains is lost through the cooling system and the exhaust gases. Present diesel engines lose about 30 percent of the energy of the fuel through the cooling system and nearly another 30 percent in the exhaust heat. Diesel engineers are trying to find ways to capture this lost energy and put it to work to increase the power and efficiency of the engine. Ceramics are an important focus of their efforts.

Researchers have studied ceramics extensively during the past 30 years as potential replacements for metal parts in the hottest region of the diesel engine, the cylinder. The metal materials that currently make up the cylinder, pistons, valves, and other parts in this hot section get too hot and fail unless the engine is cooled with water. Also, the metals conduct heat easily, so that much heat travels through the metal parts and is wasted. Some ceramics can survive higher temperature than the metals and also are poor conductors of heat. If these ceramics could be used to replace the metals in the hot section of the engine, they would require little or no water cooling and also would block the flow of heat through the walls of the engine. The heat could all be concentrated in the exhaust gases, where some of the energy could be converted to useful work by a turbocharger or other equipment.

Cummins Engine Company, in a program funded by the Army Tank and Automotive Command, evaluated replacing diesel-engine hot-section parts with ceramics. They replaced the intake manifolds, intake ports, cylinder liners, head plates, valve guides, valves, pistons, exhaust manifolds, and exhaust ports of a diesel engine with advanced ceramics—especially zirconium oxide, which can withstand high temperature and is also a good heat insulator. The engine was installed in a military vehicle and actually driven on the open road for hundreds of miles with no water cooling. Although this program showed that ceramics can work to improve diesel

Figure 9-13. Experimental ceramic components for automotive engines. (Photo courtesy of Kyocera.)

engines, the costs of the experimental ceramic parts were too high to be practical in the short term for our cars. The future is still wide open for the improvement of diesel engines with ceramics.

Ceramics in Alternate Engines.   Other types of engines are being evaluated as alternatives to gasoline and diesel engines to propel motor vehicles; the goal is cars with better than 80 miles-per-gallon fuel efficiency by about the year 2005. These alternatives include gas turbines, Stirling engines, fuel cells, and hybrid electric systems (that combine an electric motor with an engine). All of these alternatives require some ceramic parts, especially the gas turbine. In fact, companies in the United States, Europe, and Japan have been trying since the 1970s to develop a ceramic gas turbine engine, but they still have a long way to go before the engines will be able to compete with current car engines. The systems most likely to meet the 80 miles-per-gallon goal first are combinations of a gasoline or diesel engine with an electric motor.

CERAMICS IN THE PASSENGER COMPARTMENT

Some of the ceramic materials in the passenger compartment of your car are a lot more obvious than many of the ones we've talked about under the hood, but others are just as hidden. The most obvious ceramic-based material is the glass used in the windows, mirrors, and light bulbs. Hidden ceramics include the electroluminescent lighting for the instrument panel, magnets in the windshield-wiper motors and all of the power accessories (power seats, locks, and windows), the piezoelectric quartz crystal in the clock, electrical devices in the radio and tape deck, the piezoelectric buzzers for the seat-belt alarm and the anti-theft alarm, and the piezoelectric impact sensor for the air bags.

CERAMIC MATERIALS IN THE PASSENGER COMPARTMENT

Windows and mirrors

Magnets in the windshield-wiper motor

Magnets in the power seat motor

Magnets in the power window motors

Magnets in the power door locks

Magnetic head in the tape deck

Magnetic particles in the recording tape

Magnetic and piezoelectric ceramics in the loudspeakers

Electroluminescent lighting

Electrical devices in the radio

Windows and Mirrors

Special glass has been developed for car windows. The glass has been treated so that its surface is under a compressive prestress. This means that the glass is much harder to break than normal glass. The surface compressive stress also causes the glass to break in a special way. Rather than shattering into sharp spikes that could cause severe cuts, the glass breaks into little blocks that are much less dangerous. You've probably seen these pieces lying on the road after a car collision. The special glass in car windows has saved a lot of lives and lessened the severity of injuries for people who have been unfortunate enough to be in car accidents.

Mirrors also are made of glass, which is coated on the back side with a thin, shiny layer of metal. Our mirrors are another very important safety feature of our cars. Some modern car mirrors are even being made of photochromic glass, the glass discussed in Chapter 4 that darkens in bright sunlight and fades when the sun goes down.

Ceramic Magnets

You're probably surprised to learn that magnetic ceramics are used in so many different places in your car. Ceramic magnets are required in all of the electric motors for power accessories, including the windshield-wiper motors, power seat motors, window-lift motors, door locks, antenna-lift motor, defogger motor, headlight door motor, heating and air-conditioning motor, tape drive motor, and the sunroof motor. Magnetic ceramics are

AUTOMOTIVE MAGNET APPLICATIONS

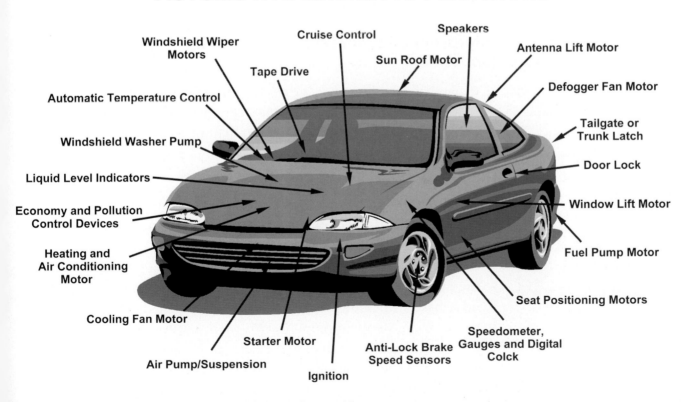

Figure 9-14. Ceramic magnets in a typical automobile. (Courtesy of Group Arnold, Marengo, IL.)

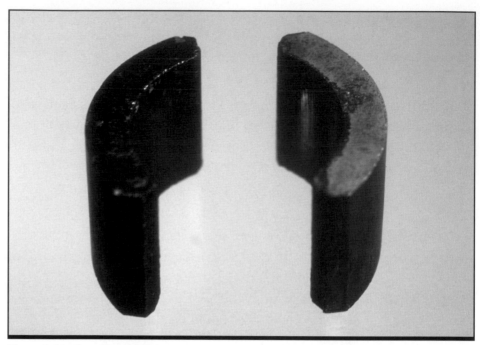

Figure 9-15. Ceramic windshield-wiper-motor magnets. *(Supplied by Delphi Chassis Systems, Division of General Motors. Photo by D. Richerson.)*

also critical in the tape deck, the speakers, the cooling-fan motor, the fuel-pump motor, the cruise control, the starter motor, and the antilock brakes.

Electroluminescent and Piezoelectric Ceramics in Automobiles

Do you remember our discussion of electroluminescent lamps in Chapter 4? EL lamps provide the lighting for the instrument panel, radio, clock, car phone, and sometimes even the overhead lighting in many cars. EL lamps are very low-cost, use little electricity, give off no heat, and last a long time with no maintenance.

Piezoelectric ceramics are used in several ways in the passenger compartment of your car. The first that you probably notice when you get into your car is the quartz clock. The second is probably the seat-belt alarm if you try to start the car without buckling up. Piezoelectric ceramics are great noisemakers when they vibrate at a frequency that's audible to our ears. Many automobile anti-theft alarms also use piezoelectric ceramics. Another piezoelectric ceramic use is in loudspeakers, along with ceramic magnets. Yet another important use is for the sensor that decides whether the air bags should be set off when your car runs into something.

amazing facts

Every automobile has at least 15 ceramic magnets.

Figure 9-16. *Electroluminescent lamp for the instrument panel of a car, shown before and after installation.* (Photo courtesy of Durel Corp., Chandler, AZ.)

CERAMICS IN THE REST OF YOUR CAR

We still aren't done finding applications for ceramics in your car. Besides windows, other parts of the auto body make use of glass. Many cars have body panels made of plastic reinforced with glass fibers. These panels are much lighter than metal panels and will never rust. Their low weight helps give new cars better gas mileage. The panels also reduce noise and vibrations and are less expensive to make. Hoods, bumper beams, fenders, roof panels, door panels, trunk lids, and spare-tire carriers all have been made out of glass-fiber-reinforced plastics.

Some composites used in cars and trucks are as fascinating as those used in airplanes. One example is an alternative drive shaft made of a mixture of glass and carbon fibers in epoxy or another advanced polymer. Metal drive shafts for long vehicles, like some trucks

> **OTHER CERAMIC USES IN YOUR CAR**
>
> Fiberglass car-body panels
>
> Composite leaf springs and drive shafts
>
> Brake parts
>
> Headlights and taillights
>
> Fuel-tank-level sensors
>
> Road-clearance sensors
>
> Rear-obstacle sensors

Figure 9-17. *Lotus Elise automobile with the body constructed of glass-fiber-reinforced polymer and brake components fabricated from a Lanxide Automotive Company ceramic-metal composite.* (Photo courtesy of Lotus Cars, Norfolk, England.)

> **CERAMICS USES IN MANUFACTURING CARS AND TRUCKS**
>
> Cutting-tool inserts
>
> Grinding wheels
>
> Sandpaper
>
> Laser cutters
>
> Welding equipment
>
> Molten-metal casting equipment
>
> Heat-treatment furnaces
>
> Sensors in robots
>
> Electronics for automation

and sport utility vehicles, have to be made in two shorter sections to avoid vibration-instability problems; they need a rigid support structure where they are attached to each other, which adds a lot of complexity and extra parts to the vehicle. A single-piece composite drive shaft that doesn't need midspan support really reduces cost and weight.

The other advanced composite, made of glass fibers in epoxy, is used for leaf springs. Leaf springs are used in many vehicles, including trailers and light trucks. They support the weight of the vehicle resting on the axles and absorb some of the shock when you hit a rut or bump. Composite leaf springs weigh up to 80 percent less than steel leaf springs.

CERAMICS IN MOTOR VEHICLE MANUFACTURING

As important as ceramics are in automobiles, they're equally important in manufacturing parts for the automobile and in robotically assembling all of these parts. Nearly every metal part in the automobile touches ceramics somewhere between the raw-material and the finished-product stages. All of the metal parts are melted in ceramic containers in ceramic-lined furnaces, which often are heated with ceramic burners. Some parts are cast into ceramic molds and machined to final shape with ceramic cutting tools. Others are hot-rolled, stamped, and welded to form the auto body. Ceramics are used extensively in the welding operation and also in the final grinding and sanding operations. We'll talk about the ways that ceramics are used in forming metals into useful shapes in the next two chapters.

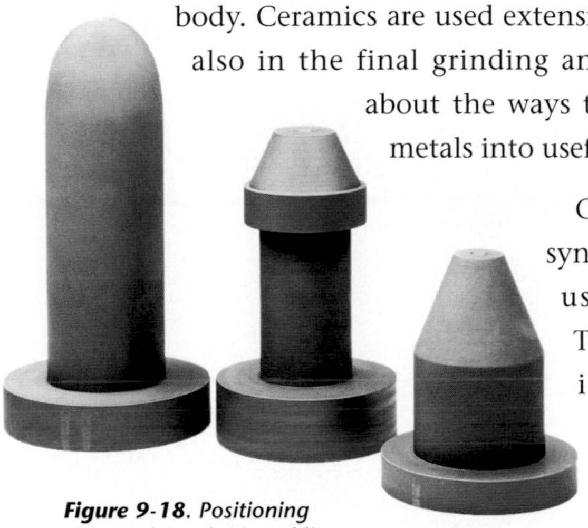

Ceramics are even important in the synthesis and weaving of the textiles used in the interior of your car. They're also necessary in the chemical process that makes gasoline from crude petroleum. These important applications of ceramics also are reviewed in the next few chapters.

Figure 9-18. Positioning pins used in robotic welding operations, fabricated from silicon nitride ceramic. (Photo courtesy of Ceradyne, Inc., Costa Mesa, CA.)

OVERVIEW: MOVING DOWN THE ROAD

Were you surprised by many of the hidden uses of ceramics in automobiles and by all of the ways ceramics are necessary in manufacturing an automobile? Today's automobiles wouldn't be possible without the many magical characteristics of ceramics. But what's down the road for automobiles? More magic! Just imagine . . . smart suspension systems that use piezoelectric ceramics to detect bumps in the road and automatically adjust the stiffness of the suspension system using other piezoelectric devices; advanced highways with cars that can navigate safely without a driver; cars with advanced-ceramic pollution-control systems that emit zero pollution; cars that run on hydrogen from water so that they're twice as efficient as present cars and emit no pollution. We live in an amazing and exciting age.

Heat Beaters

CHAPTER TEN

*J*ust about every process for converting raw materials into a useful product involves heat. You've already seen many examples of this in earlier chapters: refining, melting, and shaping of metals; fabrication of pottery and spark plugs and most other polycrystalline ceramics; melting and forming of glass; and melting of ruby and cubic zirconia to make single crystals. All of these processes need to be enclosed or surrounded with materials that can withstand these high temperatures and not react chemically with the product being processed. Ceramic materials are usually the only options.

PRODUCTS AND USES

Furnace linings for metals smelting
Furnace linings for metals refining
Burners and heaters
Kiln furniture
Crucibles
Molds and cores
Molten-metal filters
Heat exchangers
Extrusion dies

You've also learned about other places where heat is a difficult challenge, such as inside the diesel and gasoline engines that power our cars and inside the gas turbine engines that power airplanes and generate electricity. Ceramics are being

MOST CERAMICS MELT OR DECOMPOSE AT INCREDIBLY HIGH TEMPERATURE COMPARED TO OTHER MATERIALS

Material	Temperature(°F)
Polyethylene plastic	250
Aluminum metal	1220
Stainless steel	2600
Silicon dioxide (SiO_2) (Space Shuttle tiles, crucibles for melting silicon for ICs)	3000
Alumina (Al_2O_3) (weld nozzles, furnace linings, laser rods)	3720
Beryllium oxide (BeO) (laser bores, electrical insulators)	4660
Magnesium oxide (MgO) (furnace-lining material in steel making)	4750
Hafnium carbide (HfC) (rocket-nozzle linings)	6940

used more and more to meet these challenges. They're also being used to withstand the incredibly hot fireball inside rocket motors and to protect the Space Shuttle during its ascent into space and its rugged reentry into the Earth's atmosphere.

Ceramics have the unique ability to beat the heat, to survive the incredibly high temperatures that melt or otherwise degrade most other materials. This chapter reviews some of the ways that ceramics beat the heat in metals processing, industrial applications, energy conversion (power generation), and aerospace technology.

CERAMICS FOR THE REFINING AND PROCESSING OF METALS

Metals are still the dominant materials in our modern society, but without ceramics we wouldn't be able to separate the metals from their ores, refine or smelt them to a high quality, or form them economically into the endless numbers of shapes we require. Figure 10-1 lists the major steps necessary to convert metal-containing ore mined out of the Earth into a useful metal product. Let's look at the many ways that ceramics are used in each step, especially where high temperature is involved.

Raw-Materials Preparation

Metals are rarely found in the Earth's crust in their pure form. Most of the time, they're combined with other chemical elements such as oxygen, silicon, carbon, and sulfur and scattered throughout non–metal-containing rock. Preparation of raw materials involves mining the rock in which the metal is trapped, crushing and grinding the rock, and separating (concentrating) the metal-containing portion from the nonmetal portion. For most metals, high temperatures aren't involved in these crushing, grinding, and separating operations. However, the equipment is continually exposed to wear and tear caused by the chunks and particles of rock passing through the equipment. Ceramics, because they are so hard and resistant to gouging and scratching, are often used as wear-resistant tiles to line areas of the equipment that are most susceptible to wear.

METALS PROCESSING STEP	WHERE CERAMICS ARE FOUND IN EACH STEP
Raw-materials preparation	Ore handling and concentration
Smelting and refining	Furnace refractories
	Heat sources (burners, heaters)
	High-temperature bubblers
	Molten-metal filters
Melting and holding	Crucibles
	Furnace refractories
	Heat sources
	Heat exchangers
Shape forming	Crucibles, molds, cores
	Insulation for molds
	Nozzles
	Extrusion dies
	Pump parts
Finishing	Cutting and grinding tools
	Sandpaper and loose abrasives
	Cleaning and deburring equipment
Inspection	Transducers for ultrasonic nondestructive inspection
	X-ray equipment components
	Dimensional measurement instruments
Heat treatment	Furnace refractories
	Heat sources
	Setter plates and fixtures (kiln furniture)
	Heat exchangers
	Conveyors
	Circulation fans

Figure 10-1. How a metal product is produced with the help of ceramics.

Some processes use chemicals to help break down the ore and separate the metal-containing portion. Ceramics are also important here because they're very resistant to chemical attack. A good example of a necessary use of chemicals is for the processing of bauxite, the ore from which most aluminum metal is obtained. After the bauxite is crushed and ground into a fine powder, it's mixed with water and chemicals and "cooked" in a large pressure cooker, at about 500 pounds per square inch (psi) pressure and nearly 300°F. This treatment actually dissolves the aluminum-containing mineral (called *gibbsite*) out of the bauxite, just like water dissolves sugar or our stomach acid dissolves food. And much like sugar forms little crystals when the water is evaporated, the gibbsite can be separated from the water and chemicals. The final step is to heat the gibbsite to over 2000°F in a high-temperature oven (furnace) to convert it to a fine powder of aluminum oxide, which is the *feedstock* used to make aluminum. This step is where ceramics especially are required.

The furnace is a little like the oven in your kitchen but much larger, and it operates at a temperature about four times higher than your oven can reach. Because of the high temperature, the furnace must be built out of high-temperature ceramics such as alumina, zirconia, silica, or magnesium oxide. Ceramics that are used to build a furnace or other high-temperature vessel are usually referred to as *refractories*. The furnace is heated on the inside by burning a fuel such as natural gas or by using ceramic electrical heating elements similar to the heating elements inside your home electric oven. The ceramics that make up the walls, top (roof), bottom (hearth), and door of the furnace have several purposes: (1) to withstand the high temperatures on the inside of the furnace, (2) to have minimum chemical reaction with the atmosphere and materials inside the furnace, and (3) to retain the heat produced by the fuel or electricity inside the furnace. Examples of how ceramics do these three jobs are spread throughout this chapter.

Smelting and Refining

The concentrate coming out of the raw-materials preparation step is still not a pure metal. It's usually a metal or group of metals combined chemically with another chemical element, such as aluminum combined with oxygen as alumina (Al_2O_3). In the case of alumina, the aluminum and oxygen must be separated to get aluminum metal. The process for separating the metal from the other chemical elements is called *smelting*. Smelting is a high-temperature process by which the concentrate is melted, the metal is separated, and most of the remaining chemical elements are removed in the form of a molten scum called *slag*.

When you're heating soup on the stove or in the microwave oven, you need a container, or vessel, to keep the soup from flowing all over the kitchen. You also want a container that the soup doesn't stick to or eat a hole through. These requirements are the same for smelting, but the challenge is much greater because we're dealing with molten metals and slags that will react with, or eat into, just about every material imaginable at the very high temperatures necessary for smelting. Ceramic refractories are the only materials that can withstand these severe conditions, and even they are slowly eaten away.

Let's look at how ceramics are used to smelt two of our most important metals: aluminum and iron. About 10 million tons of aluminum and 95 million tons of steel are produced in the United States each year. Let's start with aluminum.

Aluminum Smelting. Aluminum requires a special smelting procedure referred to as *electrolysis* to separate the aluminum from the oxygen in the aluminum oxide feedstock. You may have heard the term electrolysis used to describe breaking down water into hydrogen and oxygen. In the case of water, the negative terminal of a battery is attached to one strip of metal and the positive terminal to another strip of metal. The two strips, which are now negatively and positively charged *electrodes,* are immersed in a container of water. If enough electricity is supplied by the battery, the hydrogen and oxygen molecules of water (H_2O) are pulled apart into negatively charged oxygen ions and positively charged hydrogen ions. We talked a little about ions in the previous chapter; an ion is an atom with an electric charge because it either has an extra electron (or electrons) or not enough electrons. Each negatively charged oxygen ion (with two extra electrons) moves to the positive electrode and combines with another oxygen ion to form oxygen gas (O_2), which bubbles to the surface of the water. Each positively charged hydrogen ion (one electron short) moves to the negative electrode and combines with another hydrogen ion to form hydrogen gas (H_2), which also bubbles to the surface of the water.

The electrolysis of aluminum oxide is similar to that of water, but is done at high temperature, instead of room temperature, in a large chamber called an *electrolytic cell.* Part of an electrolytic cell for aluminum smelting is a tank about 30 feet long, 10 feet wide, and a little over 3 feet deep. The bottom of the tank is lined with electrically conductive carbon (a ceramic) that's hooked up to the negative terminal of a direct current electricity source. This is

USES OF CERAMICS IN METALS SMELTING AND REFINING
Refractory furnace linings
Electrodes for the electrolytic smelting of aluminum
Nozzles, valves, and troughs for molten metal
Linings for oxygen lances
Molten-metal filters
Oxygen sensors
Equipment for removing dust from hot flue gases
Equipment for the reuse of waste heat

the negative electrode (called the *cathode*), where molten aluminum goes during electrolysis. The sides of the cell are also lined with carbon blocks but are not hooked up to electricity. Around the outside of the carbon electrode and side walls are layers of ceramic refractories, and around these is an outer shell, or wall, of steel. The ceramic refractories keep the heat and molten chemicals from leaking out of the electrolytic cell and protect the steel shell from the high temperature inside the cell, just like a potholder protects your hand when you take a hot casserole out of the oven.

During electrolysis, the electrolytic cell is partially filled with a molten ceramic compound called *cryolite* (made up of atoms of sodium, aluminum, and fluorine) at about 1760°F (960°C). Alumina powder that was obtained from the bauxite ore is sprinkled into the molten cryolite, where it dissolves. This molten mixture of alumina and cryolite is incredibly reactive chemically and wants to dissolve anything that it touches. Fortunately, carbon and a few other ceramic materials are resistant enough that they can be used to line the electrolytic cell. A second electrode (called the *anode*), also made of electrically conductive carbon, is connected to the positive terminal of the direct current electricity source and lowered from above into the molten cryolite–alumina, but not far enough down to touch the cathode.

When a large amount of electricity is turned on, the alumina dissolved in the cryolite is broken down into aluminum and oxygen. The positively charged aluminum ions move to the negatively charged cathode, where they form a pool of molten aluminum that can be removed from the cell by opening a valve. The negatively charged oxygen ions move to the positively charged anode, but rather than bubbling off as oxygen, they react with the carbon anode to form carbon dioxide gas. This causes two problems: First, the reaction between the oxygen and the carbon consumes (eats up) the anode, so that new anode material must continually be fed into the cell. Second, carbon dioxide is thought to be the major source of global warming by the "greenhouse effect," which you've read about in the newspapers. The aluminum industry is conducting extensive research to invent a ceramic material to replace the carbon anode, so that the anode won't have to be continuously replenished and the cell will give off oxygen rather than carbon dioxide. Finding this material has been an incredibly difficult challenge, though, because the new material must be a good conductor of electricity, withstand the high cell temperature, and not react with the molten cryolite.

As you can imagine, extracting alumina from bauxite ore and smelting the alumina to produce aluminum metal uses a lot of electrical energy and gives off a lot of pollution. In contrast, making aluminum products by recycling aluminum cans and other scrap aluminum metal uses only about 5 percent as much electrical energy as is used to smelt new aluminum and doesn't create any carbon dioxide. Fortunately, we've made a lot of progress in recent years in recycling aluminum. Presently, we recycle about 3.5 million tons of aluminum each year in the United States. And, as you can probably guess, ceramics are used in many of the steps to make recycled aluminum products.

Iron and Steel Smelting and Refining. As mentioned earlier, 95 million tons of steel were produced in the United States in 1996, but this was a small portion of the 727 million tons produced worldwide. The first step to making steel is to separate iron from iron ore. This is normally done in a *blast furnace.* A blast furnace is a tall, multistory vessel completely lined on the inside with ceramic refractories several feet thick. Iron ore concentrate, crushed limestone (a common rock made of calcium, carbon, and oxygen), and coke (carbon prepared from coal) are dumped into the top of the blast furnace, and hot air and fuel are fed into the bottom of the blast furnace. The coke burns to form carbon monoxide, which reacts chemically with the iron ore to produce molten metallic iron that's about 2840°F (1560°C) in temperature. Other materials in the ore react with the limestone to form a molten slag that floats on top of the iron because it's lighter than the iron. The molten iron and molten slag are removed from the blast furnace through high-temperature valves and nozzles (sort of like large water taps).

> **ALUMINUM-SMELTING FACTS**
>
> Aluminum metal is produced from aluminum oxide ceramic powder.
>
> Smelting vessels are lined with ceramics.
>
> Smelting consumes a great deal of electricity and gives off carbon dioxide gas.
>
> Recycling uses 95 percent less energy and eliminates carbon dioxide emissions.
>
> About 60 percent of aluminum beverage cans were recycled in 1990.

Surviving the continuous flow of solid raw materials, liquid metal and slag, and hot gases through the blast furnace is a difficult challenge for the ceramic refractories lining the inside of the furnace. Remarkably, ceramics have been developed that can survive for six to seven years before the blast furnace must be shut down and relined.

Most iron that leaves the blast furnace contains about 3 to 5 percent carbon, which is okay for the cast iron in the engine of our cars but not for the many different grades of steel required for other parts of our cars, or for buildings, bridges, and many other products that we encounter every day. Most of this iron is processed in a *basic oxygen furnace* (BOF). The conditions in a BOF are at least as severe as those in a blast furnace, so the inside must be completely lined with ceramic refractories. Rather than working continuously, the BOF is operated in cycles that last about one hour each. The first step of each cycle is to load the BOF with about 100 tons of scrap steel, 250 tons of 2370°F (1300°C) molten crude iron, fluxes (materials that react with impurities to form slag), and other metals (needed to form the desired steel alloy). Loading all of that material into the furnace exposes the ceramic lining to heavy impact and abrasion from the scrap steel and to severe thermal shock from the molten crude iron.

The second step in the BOF cycle is to force a large, hollow tube, called a *lance,* from the top of the furnace into the mixture of iron, fluxes, and scrap steel. Pure oxygen then is pumped through the lance. The oxygen bubbles through the molten iron and begins to react with the carbon.

This reaction gives off heat, and so the temperature of the furnace increases to between 2900° and 3100°F. The scrap steel melts, the fluxes react with impurities and form a slag that floats on the surface, and the metals mix to form the desired steel alloy. All through the cycle, the ceramic lining is bathed in churning molten steel and molten slag. The lining ranges from about 18 to 36 inches in thickness and is expected to survive thousands of cycles. It's amazing that any material can withstand such severe conditions for even a single cycle, but ceramics have met the challenge and survived for thousands of cycles.

Although the molten metal from smelting or refining is sometimes formed immediately into a product, it's usually just cast into simple, temporary shapes called *ingots* for later remelting or forming by other methods. Casting a molten metal into an ingot or product is a critical step. Often the molten metal contains little chunks of debris picked up from the refractories or present in the original scrap steel. These chunks contaminate the steel and can cause a part to be rejected or fail during use. During the 1980s, engineers learned to make special porous ceramic filters that could survive the high temperatures of the molten metals and screen out the debris as the metal was being poured into ingots or products.

One of the most interesting and effective ceramic filters for "straining" molten metal into a casting mold is produced using a sponge similar to the kind you may have in your kitchen. A ceramic slip (a suspension of ceramic particles in a liquid), similar to those used for slip casting beer mugs and figurines, is poured into a sponge until its pores are completely filled. This ceramic-filled sponge is carefully dried in an oven. It's then fired in a furnace at a high temperature to burn out the sponge and bond the individual ceramic particles into a solid ceramic, just like a piece of ceramic pottery or an alumina spark-plug

insulator is fired. The resulting part looks like a continuous network of interconnected ceramic branches plus a continuous network of pore channels. Examples are shown in Figures 10-2 through 10-4. This skeletal structure is very resistant to thermal shock and doesn't block the flow of molten metal, but it does effectively filter out any debris in the metal.

Ceramic molten-metal filters were introduced about 20 years ago and have become standard for filtering almost all molten metals. I recently had the opportunity to tour a large cast-iron foundry where Caterpillar produces diesel-engine castings for trucks. All metal cast in that foundry passes through ceramic filters. The tour guide, who had been at the foundry for 35 years, stated that ceramic filters were the most important innovation during the past 20 years for improving alloy quality and reducing faulty castings.

Ceramics certainly are necessary for smelting and refining metals. Refractories, electrodes, tap valves, and molten-metal filters all are important, but ceramics also are used in other ways, such as for transferring heat from exhaust gases to preheat air going into the process (heat

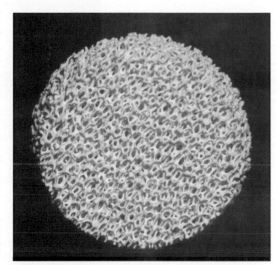

Figure 10-2. *Ceramic molten-metal filter, showing the sponge cell structure.* (Sample from Hi-Tech Ceramics, Alfred, NY. Photo by D. Richerson.)

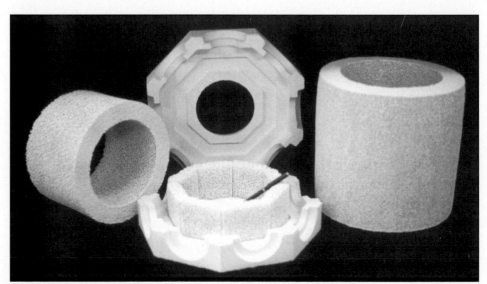

Figure 10-3. *Assembly of ceramic refractories and a ceramic molten-metal filter.* (Photo courtesy of Vesuvius Hi-Tech Ceramics, Alfred, NY.)

Figure 10-4. *Examples of various configurations and compositions of ceramic molten-metal filters.*
(Photo courtesy of Vesuvius Hi-Tech Ceramics, Alfred, NY.)

exchange), helping remove dust particles from exhaust gases, and sensing the amount of oxygen in molten steel. Reuse of waste heat and cleanup of particles drawn along with the gases up the stack (flue) are important challenges in several steps of metals processing and in many other industrial processes. These jobs are discussed later in this chapter and in subsequent chapters.

Remember the zirconia oxygen sensors that we discussed for cars? That same technology is used to measure the oxygen content in molten steel. Before the development of the oxygen sensor, a steelworker had to scoop a sample of molten steel out of the furnace and send it to the laboratory for analysis. That process took about 20 minutes. Now, a ceramic oxygen sensor can be thrust directly into the molten steel to determine the oxygen content within a few seconds. This innovation has both improved the control and speeded up the process of steelmaking.

> **CERAMICS FOR METAL MELTING AND HOLDING**
>
> Burners
>
> Furnace linings
>
> Heat exchangers
>
> Crucibles
>
> Thermocouple protection tubes

Melting and Holding

Casting is a common method for forming smelted and refined metals into complex shapes. Sometimes, the molten metal is poured directly through a ceramic filter into a shaped ceramic mold to cool and become a solid metal part. Other times, metal ingots are remelted in a ceramic-lined container (such as a crucible) inside a ceramic-lined furnace and then often transferred to a separate, ceramic-lined furnace to be held at temperature until the molten metal is ready to cast. Many melting and holding furnaces are heated by natural gas, with ceramic *radiant burners*: A natural gas and air mixture is burned inside a ceramic tube, and the heat radiates from the tube (like the heat radiates from a fireplace or the sun) onto the surface of the metal being melted. This type of heating minimizes contamination of the metal.

Only a portion of the heat produced by a radiant burner or other heat source is captured to melt the metal. About two-thirds of the heat goes up the chimney or smokestack and is wasted. Some of this wasted heat can be reclaimed through the use of a *heat exchanger*. Most industrial heat exchangers consist of pipes or tubes that line or crisscross the chimney, the stack, or a chamber that holds the hot gases before they reach the chimney or stack. Cold air on its way to mix with the natural gas fuel is pumped through the heat-exchanger tubes, where it picks up heat from the exhaust gases. The more heat that can be picked up, the less this air needs to be heated by the burning fuel to reach a temperature high enough to melt the metal—and the less fuel is consumed. Ceramic heat-exchanger tubes can reduce the amount of fuel needed by about 50 percent, a tremendous savings.

Other ceramics are important for melting and holding furnaces. One ceramic application is for protection tubes to separate the thermocouples

Figure 10-5. Silicon carbide radiant burner/heat exchanger tube sections. (Photo courtesy of Schunk-INEX Corp., Holland, NY.)

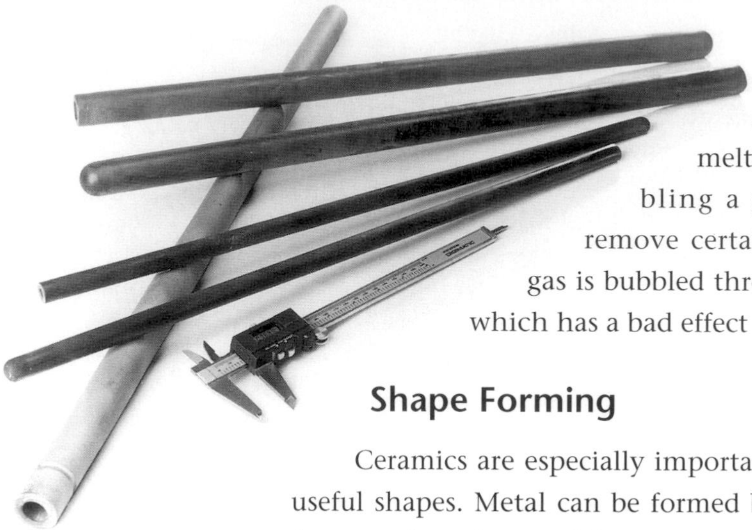

Figure 10-6. Silicon nitride thermocouple protection tubes. (Photo courtesy of Ceradyne, Inc., Costa Mesa, CA.)

(metallic temperature-measurement devices) from the molten metal. Another use is associated with further metal refining sometimes conducted in the melting or holding furnaces, such as bubbling a gas through the molten metal to remove certain impurities. For example, chlorine gas is bubbled through aluminum to remove hydrogen, which has a bad effect on the properties of the aluminum.

Shape Forming

Ceramics are especially important for forming metals into their final, useful shapes. Metal can be formed by many ways into the shape needed for a product. Some of those ways include casting from a melt, extruding, rolling, forging, and cutting/grinding. Let's talk about casting first, because that process uses ceramics extensively and is one of the most cost-effective approaches for making complex shapes such as automobile engines, industrial equipment, parts for household appliances, and gas turbine engine components.

Ceramics are necessary in just about every step of casting. The container, or crucible, that holds the molten metal prior to casting is ceramic. Pouring spouts and troughs that guide the metal from the holding furnace or crucible to the shaped casting mold are lined with ceramics. Thermocouples that measure the temperature of the molten metal are enclosed in protective ceramic tubing. The molten metal is poured through a ceramic filter immediately before it flows into the mold.

Many methods have been invented over the centuries to cast metals. Most of them involve the use of a ceramic mold. The mold is the shaped cavity into which the molten metal is poured. The mold must be strong enough to hold the weight of the metal, must not shatter from thermal shock when the extremely hot molten metal is poured into it, must not react chemically with the hot metal, and must be easy to remove after the metal has cooled and become solid.

Lost Wax Casting.   One particularly interesting casting process that uses ceramics extensively is *investment casting,* also known as *lost wax casting.* Lost wax casting starts by creating a wax model of the part to be cast. The wax model usually is produced by melting wax, pouring or injecting it with low pressure into a shaped metal cavity, and allowing the wax to cool and solidify. The metal cavity is constructed of multiple pieces, so that it can be opened to remove the wax model in one piece. The wax model is dipped into a slip (slurry) of ceramic particles suspended in water. The slip has the consistency of paint. A layer of this ceramic "paint" sticks to the wax, to form a thin layer of ceramic particles, after the water has evaporated. The coated wax is dipped again, but instead of being dried, it is immediately coated with larger ceramic particles about the size of sand. This process of dipping in the slip and then coating with ceramic sand is repeated about six or seven times to build up a layered ceramic coating about one-eighth to one-quarter of an inch thick around the wax. The wax is then melted and allowed to flow out of the ceramic shell. The ceramic shell is now a mold that has the exact shape inside as the outer shape of the wax model. The ceramic mold is dried and fired at a high enough temperature that the ceramic particles bond together, giving the mold adequate strength to not break when it's later filled with molten metal.

> **CERAMICS CRITICAL TO SHAPING METALS AT HIGH TEMPERATURE**
>
> Furnace linings and burners
>
> Crucibles
>
> Pouring spouts
>
> Troughs
>
> Thermocouple protection tubes
>
> Molds and cores
>
> Molten-metal filters
>
> Tundish nozzles
>
> Thermal insulation
>
> Extrusion die liners

Once created, the ceramic casting mold is wrapped with ceramic-fiber insulation, heated, and filled with molten metal to produce a solid metal casting. In some cases, a casting with internal holes and passageways is desired, rather than a solid metal casting. A good example is a rotor blade for a gas turbine engine, which has a complex array of internal cooling passages. A ceramic *core* is used to form these complex passages directly during casting. The ceramic core is formed separately and then placed in the mold before the metal is poured in. The metal flows into the mold and around the core. The core is made of a special mixture of ceramics that can withstand the temperature of casting but can later be removed from the solid metal blade by dissolving with a weak acid that doesn't attack the metal. Figure 10-7 shows a mold with two blade chambers cut open to reveal the cores in place before casting. The resulting internal passages permit cooling of the turbine blades with forced air when the engine is operating, allowing the engine to run at a much higher temperature than the metal can normally tolerate. The result is higher engine power and much lower fuel consumption.

The first cooled turbine blades were achieved by an expensive process of drilling holes in the blade after it was cast. Ceramic cores make it possible to form the air passages directly and easily during casting. Core-formed passages can be much more complex than drilled-in channels, and so can lead to more efficient cooling and much greater increases in engine power and reductions in fuel consumption. The ceramic-core technology has been a big factor in helping keep airline fares affordable.

Ceramics in Steel Casting. Whereas investment casting is done in batches typically well under one ton of molten metal, iron and steel casting is done either continuously into ingot molds or in large, multi-ton batches, directly from large molten-metal containers called *ladles* or *tundishes*. The bottom of the container has a valve that can be opened and closed like a water tap, except that this valve must withstand very high temperatures. For example, molten metal flows from a tundish through a *tundish nozzle* that has an opening about one-half to three-quarters of an inch in diameter. Depending on the metal alloy, the metal flowing through the tundish nozzle ranges in temperature between 2732° and 3092°F (1500° and 1700°C). Various compositions of zirconium oxide ceramic have survived this severe application.

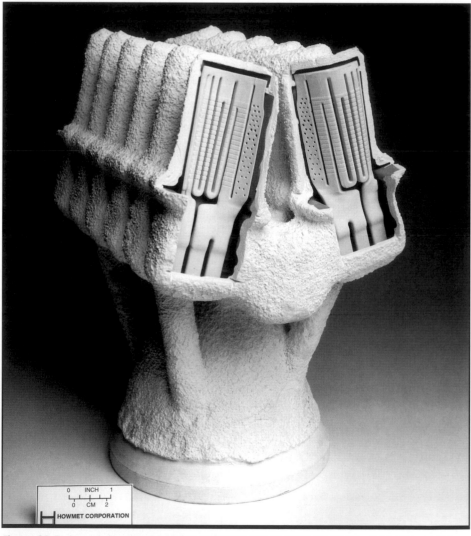

Figure 10-7. *Ceramic investment-casting mold with core in place, for casting gas turbine engine rotor blades containing intricate internal cooling passages.* (Photo courtesy of Howmet Research Corp., Whitehall, MI.)

Ceramics for Extrusion.   Ceramics are important in other metal shape-forming operations such as *extrusion.* You use extrusion every day when you squeeze toothpaste out of the tube: You press on the tube, and toothpaste "extrudes" out of the tube in the circular shape of the opening of the tube. You use the same concept when you decorate a cake, except that you can change the shape of the opening through which you squeeze the frosting to get different shapes. The same is done with metals, at a temperature that's high enough to soften the metal but below the melting temperature. The heated metal is pushed, with high pressure, through a shaped opening (*extrusion die*) at the end of the extrusion equipment. Iron-based alloys are extruded at very high temperatures (around 2200° to 2300°F) and pressures (around 100,000 to 150,000 psi). Brass and copper are extruded at lower temperatures. Solid rods and tubing in circular, square, and rectangular cross sections commonly are produced by extrusion.

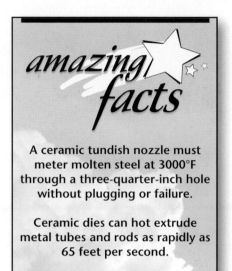

A ceramic tundish nozzle must meter molten steel at 3000°F through a three-quarter-inch hole without plugging or failure.

Ceramic dies can hot extrude metal tubes and rods as rapidly as 65 feet per second.

Extrusion can be very inexpensive. The biggest concern is wear and deformation of the extrusion die as the metal is forced through at high temperature and pressure. If the size or shape of the opening of the die changes, the die can no longer be used; metal extrusion dies must be replaced frequently. Transformation-toughened zirconia (TTZ) ceramic dies have much longer life than metal dies for some extrusion processes. For example, one TTZ extrusion die survived more than 6000 extrusions of brass, while the previous metal dies had survived only 10 to 50 extrusions. Each extrusion consisted of forcing a solid *billet* of brass about 12 inches in diameter and 30 inches long through a die opening three-quarters of an inch in diameter, at 1650°F to produce about 65 feet of brass rod each second. You can see how extrusion can be cost-effective, especially if the extrusion die survives for many extrusion runs.

Finishing and Inspection

Metal parts that come out of a casting mold, an extrusion die, or other metals-forming device generally don't have the exact dimensions or surface smoothness needed for the intended

Figure 10-8. *Zirconia die liner for hot extrusion of brass.* (Photo by D. Richerson.)

Figure 10-9. *Silicon nitride rolls for hot-forming rods and tubing of non-iron metals. (Photo courtesy of Ceradyne, Inc., Costa Mesa, CA.)*

application. These parts require a finishing operation such as cutting, grinding, or polishing. Ceramics are used extensively for finishing metals. Silicon nitride cutting-tool inserts were discussed in Chapter 5, and other ceramic cutting-tool inserts are discussed in Chapter 11. A cutting-tool insert isn't normally thought of as a high-temperature use for materials, but it really is. When metals are shaped by machining, the ceramic cutting tool can reach temperatures greater than 2000°F. In fact, the high-temperature stability of the ceramics is what gives them a competitive edge over metallic cutting tools. Other ceramic finishing products include grinding wheels, sandpaper, and loose abrasives.

Ceramics are also important components in inspection tools for metals, but these uses don't involve high temperature. Examples of such applications include the piezoelectric transducer for ultrasonic nondestructive examination that we talked about in Chapter 7, bearings and other components in X-ray tubes also used for nondestructive inspection, and wear-resistant surfaces on measurement instruments.

Heat Treatment

Most metals do not have optimum properties immediately after casting, extrusion, or other shape-forming processes. They usually require at least one additional high-temperature treatment. For example, extrusion and hot rolling (to make sheet metal and foil) involve severe deformation of the metal, which builds up stresses. These stresses can be eliminated by aging or annealing the metal (holding it for a while at an elevated temperature). Sometimes, the metal needs to be held at a high enough temperature for the microstructure to change. Other times, the end use requires that the surface of the metal be harder than the interior, and so the part is placed in a hot furnace containing carbon or nitrogen that reacts with and hardens the surface. Sometimes, the metal must be heated and then cooled very quickly (*quenched*). All of these treatments are categorized under the general term *heat treating.*

> **USES OF CERAMICS FOR HEAT-TREATING METALS**
>
> **Furnace linings**
>
> **Burners and heating elements**
>
> **Kiln furniture**
>
> **Heat exchangers**
>
> **Thermocouple protection tubes**
>
> **Conveyor belts**

Figure 10-10. *Silicon carbide segments of a conveyor belt for a powder metal sintering furnace.*
(Samples courtesy of Saint-Gobain/Carborundum Structural Ceramics, Niagara Falls, NY, photo by D. Richerson.)

Heat-treating temperatures can vary anywhere from a few hundred degrees to around 2000°F, depending on the metal involved and the objective of the heat treatment. The ceramic needs are the same as in other high-temperature metal-processing steps: furnace linings, heat sources such as natural gas burners and electrical heating elements, thermocouple protection tubes, and heat exchangers. However, the temperatures for heat treatment are lower than for handling molten metal. Since the metals aren't molten, crucibles and molds aren't required. Instead, the metal parts require a solid surface to rest on. These surfaces usually are made of ceramics and are called *kiln furniture*. Other

Figure 10-11. *High-temperature chain fabricated from aluminum oxide ceramic, by Ceramic Binder Systems, Butte, MT.* (Photo by D. Richerson.)

ways in which ceramics are sometimes used for heat treating are for conveyor belts to carry metal parts through a tunnel-shaped furnace, and as fan blades to circulate the hot gases inside an enclosed furnace, to make sure all of the metal parts are exposed to the same temperature and furnace atmosphere. Figure 10-10 shows several small silicon carbide segments of a high-temperature conveyor belt that is 18 inches wide and 120 feet long when fully assembled.

CERAMICS FOR HIGH-TEMPERATURE INDUSTRIAL PROCESSES

Many of the ways ceramics are used in industrial processes are similar to those in metals processing: for furnace linings, high-temperature fixtures, burners, and heat exchangers. These parts are required in glass production, ceramics production, chemicals processing, petroleum refining, and even papermaking.

Glass Production

Glass production is a particularly impressive example of how ceramics are necessary in industry. As you may remember from Chapter 4, molten glass is a universal solvent that will dissolve almost anything it touches. The large chamber (glass tank) in which glass is melted must be lined with a very stable, inert material that either is not dissolved by the molten glass or is dissolved very slowly. The metals platinum, iridium, and rhodium are extremely chemically resistant to most molten glass compositions but are too expensive for lining a large glass tank for making window glass or glass containers. For example, the furnace for making window glass can have a glass-melting tank larger than a tennis court (covering around 5000 square feet of floor space) and produce more than 25,000 tons of glass per day. These large glass tanks are lined with solid blocks of zirconia or mixtures of zirconia, alumina, and silica. The blocks look a little like the large blocks of rock that are quarried to build buildings, but they're made by actually melting the ceramic at incredibly high temperature [above 3632°F (2000°C)] and pouring (casting) the molten ceramic into a mold made of sand, graphite, or water-cooled metal.

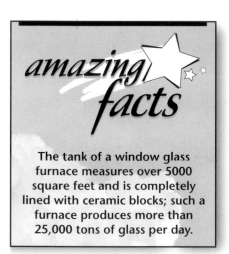

Industrial Furnaces

Many industrial processes require a heat source at some stage. Some examples include paint drying and window shaping in the automotive industry, food processing, paper drying, preparation of cement and plaster, and disposal of hazardous materials. Ceramics are involved in several different ways. In some cases, the heat is produced by burning a fuel. The chamber in which the fuel is burned generally is lined with ceramics or made completely of ceramics. A good example is the radiant burner, discussed earlier in this chapter, for melting aluminum. Another type of ceramic burner is a *radiant surface burner*. This burner uses a porous block, or plate, of ceramic. The fuel and air are mixed on the back of the porous plate, pass through the pores of the plate, and burn with a very uniform flame as they leave the plate. This type of burner produces very little pollution and has even been used in a fast-food restaurant for cooking hamburgers.

Other heat sources take advantage of the electrical characteristics of some ceramics. As you learned in Chapter 6, some ceramics let electricity pass through, but only with difficulty. Because the electricity has to work to get through, much of the electrical energy is converted to heat energy, so the ceramic gets very hot. Because no fuel is being burned, this electrical source of heating is very clean and doesn't contaminate the materials being processed or release any pollution. Silicon carbide (with additives to give the right electrical resistance), zirconium oxide, and molybdenum disilicide are important ceramics that can be used in air for electric heating. Carbon, such as graphite, is also an effective ceramic for electric heating, but carbon is attacked by air at high temperature and can only be used in a vacuum or in nitrogen, argon, or helium gas.

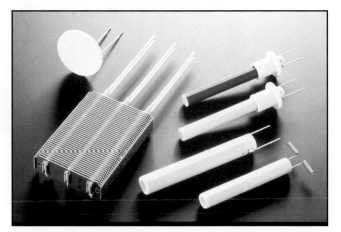

Figure 10-12. Ceramic heaters. *(Photo courtesy of Kyocera.)*

Chemicals and Petroleum Processing

Ceramics are important in many ways to chemicals and petroleum processing, two of the largest industries in the world. In these huge industries, ceramics are chosen mostly for their wear resistance and corrosion resistance, properties discussed in the next chapter; but many uses also require materials that are chemically inert at high temperatures, so again ceramics fit the bill. The largest uses of ceramics in chemical processing are for catalyst supports and heat exchangers.

As you probably remember from our discussion of the automotive catalytic converter in Chapter 9, a catalyst is a material that stimulates a chemical reaction to take

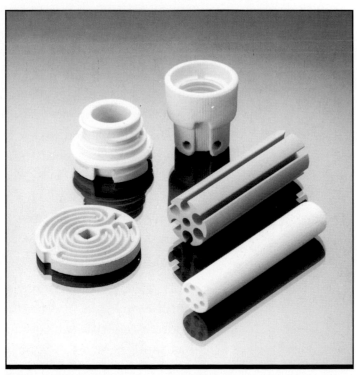

Figure 10-13. Ceramic insulators for heaters. *(Photo courtesy of Rauschert Industries, Inc., Madisonville, TN.)*

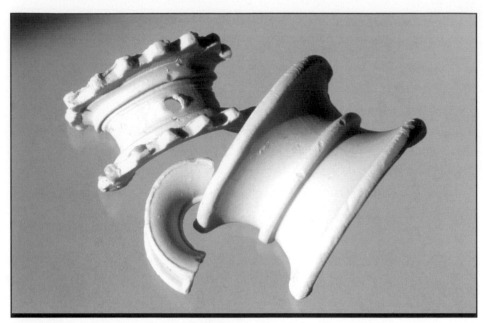

Figure 10-14. *Ceramic "saddles" for chemical processing, designed to provide temperature uniformity and controlled flow of chemicals.* (Samples supplied by Saint-Gobain/Norton Chemical Process Products, Solon, OH. Photo by D. Richerson.)

place but is not a part of the reaction. Sometimes, a heated ceramic with lots of porosity and surface area that chemicals can touch acts as a catalyst by giving the chemicals a place to react. Other times, the pores and surfaces of the ceramic can be coated with a thin layer of a catalyst such as platinum, so that the ceramic only acts as a support structure for the catalyst. A ceramic honeycomb structure acts as the catalyst support for the automotive catalytic converter. The ceramic catalyst support structures for chemicals and petroleum processing are much simpler. They're usually just small, highly porous pellets of a special alumina-based ceramic. These pellets are packed into a heated pipe, chamber, or container that the chemicals are pumped through. As the chemicals pass through the pipe, the catalyst and heat cause them to react, decompose, or distill. The new chemicals formed by the reaction come out the other end. Ceramic catalyst supports are used to break down crude petroleum into gasoline and other chemicals that are then used to make many products, such as plastics, synthetic fibers for clothing, and industrial chemicals.

Heat exchangers in the chemicals industry have the same purpose as heat exchangers in other industries: to salvage and reuse heat that would otherwise be wasted. For most chemicals processes, pieces of ceramics such as those shown in Figure 10-14 absorb heat during one portion of the process and then give back the heat in another portion of the process. This conserves fuel and energy and helps keep the cost of the process affordable.

Ceramics in the Ceramics Industry

The ceramics industry is one of the largest customers of high-temperature ceramics. It seems as if every ceramic fabrication process we've discussed is done at high temperature: firing of pottery, firing of alumina spark-plug insulators, separation of alumina powder from bauxite, growth of cubic zirconia crystals for jewelry, and synthesis of silicon nitride and silicon carbide. All of these processes require furnaces lined with ceramics and ceramic kiln furniture to support the ceramic parts being fired in the furnaces. Many require ceramic heaters, heat exchangers, and thermocouple protection tubes.

One of the biggest advances in ceramics for ceramic processing has been in the refractories that line the furnaces. Before about 1960, furnaces were lined with ceramic refractory bricks that looked similar to the bricks used for building houses. In these furnaces, the ceramic brick absorbed heat while the product in the furnace was being heated. When a furnace constructed with these bricks was turned on, the bricks had to heat up at the same time that the product in the furnace was heating. This process was slow, and a lot of the energy in the fuel was wasted to heat the bricks.

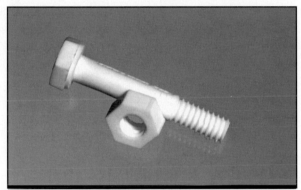

Figure 10-15. *Ceramic nut and bolt for high-temperature fastening, such as holding high temperature ceramic insulation in place inside a furnace.* (Samples from Coors Ceramics, Golden, CO. Photo by D. Richerson.)

Also, after the firing cycle was completed, the bricks contained so much heat that the furnaces cooled very slowly. The result was a long cycle time to fire one batch of ceramic product.

Since about 1960, researchers have learned to make fibers from high-temperature ceramics such as zirconia and alumina and to use these fibers as the walls of furnaces in place of most of the brick. Some of these fibers can withstand temperatures as high as 4000°F and insulate better than brick. The fibers are woven into cloth, compacted into porous fiberboard, or packed into loose bundles like the fiberglass insulation used in the walls of houses. The fibers are much less bulky than the brick and absorb less heat. Therefore, the furnace can be heated and cooled much more rapidly than with brick. This speedy heating saves a lot of fuel and greatly decreases the cycle time to fire a batch of ceramic product. As a further benefit, the fibrous insulation is flexible (rather than rigid, like brick) and easy to install and repair or replace.

Besides being a major innovation in the ceramics industry, ceramic-fiber linings have also been implemented in many of the other furnaces that we've discussed in this chapter, resulting in enormous savings for a broad cross section of industries.

ENERGY CONVERSION

Energy conversion is the process of extracting energy from sources such as coal, oil, wood, wind, water, and sunshine and converting it to a form such as electricity that we can use to heat and light our homes, run our appliances, and do all of the other things we take for granted in our modern society. Many of these processes, such as the burning of coal, oil, and wood, involve high temperatures and thus benefit from ceramics. Most of the processes we use right now to convert energy are wasteful and produce dangerous pollution. Ceramics have the potential to dramatically improve the efficiency of energy conversion and reduce the pollutants. Energy production and pollution control are so important to our modern world that a whole chapter (Chapter 12) is devoted to the ways ceramics are contributing to these tasks.

AEROSPACE USES FOR CERAMICS

The Space Shuttle

Our journeys into the sky, and even into outer space, require the services of ceramic materials in many different ways, but especially for resistance to high temperature. Probably the most familiar use of ceramics for aerospace technology is in the Space Shuttle. The outer surface of the Space Shuttle is covered with about 33,000 ceramic tiles that protect the underlying aluminum structure and the astronauts from getting too hot during ascent into space and reentry into the Earth's atmosphere. The friction of the Shuttle traveling at high speed through the atmosphere produces temperatures up to about 2650°F, about double the melting temperature of the aluminum alloys from which most of the Shuttle is built.

> **SILICA SPACE SHUTTLE TILE STATISTICS**
>
> Must protect aluminum from temperatures up to 2300°F
>
> Are made of porous silicon dioxide ceramic
>
> Are so lightweight that a cube 1 foot on each edge weighs less than 9 pounds
>
> Vary in thickness from less than 1 inch to over 4 inches

Two types of carefully engineered ceramics protect the Shuttle. The tip of the nose and the front edges of the wings, which are exposed to the highest temperatures, are made from a ceramic composite of carbon fibers in a carbon matrix. This composite has extremely high temperature resistance but burns if it comes in contact with the oxygen in the air at high temperature. To protect against contact with oxygen, the carbon–carbon composite is first coated with a

Figure 10-16. *Space Shuttle during liftoff.* (Photo courtesy of NASA.)

layer of silicon carbide and then overcoated with a layer of silicon dioxide.

Most of the Space Shuttle surface doesn't get any hotter than 2300°F and so can be protected with tiles made of silicon dioxide (silica). You already learned at the beginning of this book that silica is one of the most common materials in the Earth's crust, often found as beach sand. Silica melts above 3100°F, so it has adequate temperature resistance; but temperature resistance alone isn't enough. We can't just coat the Space Shuttle with sand, or even with blocks of silica. Besides having high-temperature resistance, the protective material must be engineered so that it's very lightweight, can withstand severe thermal shock, can actually block the flow of heat, and can withstand the erosion caused by moving at high speed through the atmosphere.

Many years of research and development were needed to achieve a silica material that meets all of the criteria for the Space Shuttle. This material is made starting with tiny fibers of silica that are suspended in water and poured into a porous mold. As water is sucked out through the pores of the mold, the fibers intertwine to form a structure that holds its shape but contains over 80 percent open space. When this structure is dried and fired at a high temperature, the fibers bond together where they contact each other to give the material a little strength, but all of the open space remains, as pores filled with air. The material looks a bit like the fine-grained styrofoam you can buy at the craft store. A cube of this Space Shuttle tile material measuring 1 foot on each edge weighs less than 9 pounds.

The open pores filled with air make the material lightweight but also are an effective barrier to the flow of heat. Heat doesn't pass easily through this dead air space or through the silica ceramic. In fact, one surface of a piece of this porous silica only 1 inch thick can be heated with a torch until it's red-hot, and you can still put your hand on the other side (only an inch away) and not feel the heat!

The material still isn't ready to be attached to the surface of the Space Shuttle. It must be cut to the right size and have a glassy surface coating deposited on it to protect against erosion and also to keep water (from rain) from soaking into the pores. If the pores picked up rain water, the Space Shuttle would be too heavy to take off. The coating also has another purpose. It's specially designed to re-radiate (give off) heat into the atmosphere almost as rapidly as the heat is produced by friction during ascent and reentry. Otherwise, the heat would soak slowly through the porous silica tile into the aluminum and would endanger the lives of the astronauts inside the Space Shuttle.

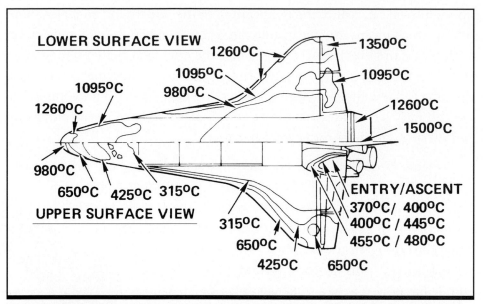

Figure 10-17. *Temperatures at various places on the surface of the Space Shuttle during ascent and reentry.* (Source: Bulletin of the American Ceramic Society, Vol. 60, No. 11 (1981), p. 1188.)

As shown in Figure 10-17, different regions of the Space Shuttle surface heat to different temperatures. Engineers have used computers to calculate the exact thickness of ceramic tile that is necessary to protect against each temperature. The thicknesses range from two-tenths of an inch to four and one-half inches. Each of the 33,000 tiles is different and is given a code number. If a tile is damaged during a mission, its number is fed into a computer, and a replacement tile is automatically carved from a new block of material.

Ceramics are used in many other ways in the Space Shuttle. Some ways are high-temperature applications, such as rocket-nozzle liners for thrusters that allow the Space Shuttle to be maneuvered while in orbit. Others include the advanced electronics, the pressure and temperature sensors, the windows, and even reinforcement for some metal parts.

Other Aerospace Uses for High-Temperature Ceramics

The ceramic tiles covering the Space Shuttle act as a thermal barrier. Ceramics are used as much thinner *thermal barrier coatings* in gas turbine engines to protect underlying metal from the high temperatures inside the engine. The most common thermal barrier coating is zirconia. A layer only about 25 thousandths of

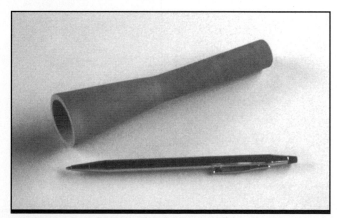

Figure 10-18. *Ceramic rocket guidance nozzle.* (Sample from Coors Ceramics, Golden, CO. Photo by D. Richerson.)

an inch thick can reduce the temperature of the underlying metal by about 400°F, which means that the engine can be run at much hotter temperatures without destroying the metal. Operating the engine at higher temperature increases the power of the engine and decreases the amount of fuel used. Zirconia ceramic thermal-barrier coatings are now standard for commercial and military jet engines and are at an advanced stage of development for industrial gas turbines for power generation.

Figure 10-19 illustrates another high-temperature use for ceramics in aircraft engines. Aircraft engines need a spark plug or igniter to light the fuel, just as a car engine needs a spark plug, but the temperature in the aircraft engine is usually much higher. The spark-plug insulator must withstand these high temperatures, plus severe thermal shock. Silicon nitride has proved to be a suitable insulator material.

Ceramics are important in other high-temperature aircraft applications. Ceramic bearings for butterfly valves and seal runners were mentioned in Chapter 5. Other ceramic and ceramic-matrix composite parts are under development for afterburner components in high-speed military aircraft, for the combustion system in the NASA supersonic transport, and for many parts in gas turbine engines.

Figure 10-19. *Silicon nitride aircraft engine igniter insulator.* (Sample from Ceradyne, Inc., Costa Mesa, CA. Photo by D. Richerson.)

Figure 10-20. *Various silicon nitride parts for aerospace and other high-temperature uses.* (Photo courtesy of AlliedSignal Ceramic Components, Torrance, CA.)

OVERVIEW: HARNESSING HEAT

Heat can be our worst enemy—or it can be our best friend, if we are able to harness it. Ceramics help to harness heat, to convert worthless ores into valuable metals, to generate power, to process chemicals and other products that we use every day, and to provide transportation on the Earth and into outer space. The magical resistance of ceramics to heat will take us places we've never been and help us do tasks that will lead our civilization to new heights.

The Hardest Materials in the Universe

Hardness represents the ability of a material to resist deformation, gouging, scratching, abrasion, and erosion—in other words, to resist wear. Wear is a major cause of failure for parts in industrial equipment, automobiles, and aircraft. A government study in the 1970s reported that wear was costing the manufacturing industry alone over $25 billion every year in maintenance and replacement. The cost is undoubtedly much higher now. Guess who ultimately pays the costs of wear? Right, we consumers. How do we combat wear? With hard materials; and ceramic materials are the hardest materials known.

PRODUCTS AND USES

Liners
Seals
Pump parts
Oil-recovery equipment
Cutting tools
Chemical-processing equipment
Papermaking equipment
Coal-handling equipment
Wire-drawing tooling
Thread guides
Sandblast nozzles
Ore-processing equipment
Ceramics-processing equipment
Can-forming tooling
Magnetic storage disks
Bearings

Not all ceramics are hard, but those that are hard are extremely hard. To understand the definition of "extremely hard," we need first to learn how hardness is measured and then to compare the hardness of various materials. The hardness of a material usually is measured by pressing the pointed end of a pyramid-shaped indentor made of diamond against the polished surface of the material and determining the amount of pressure (load) that's necessary to produce a dent in that surface. The size of the dent is then measured, so that the amount of pounds per square inch or kilograms per square millimeter of force required to cause the dent can be calculated.

Diamond is the hardest of all known materials, which is the reason indentors are made of diamond. A force of over 10,000,000 pounds per square inch (7000 kilograms per square meter) is necessary to dent a diamond! That's equal to stacking about 800 six-ton elephants on a postage stamp. The hardnesses of some other ceramics are listed in Figure 11-1, which you can refer to as we discuss various hard ceramics and their uses throughout this chapter. For comparison, most metals indent well below about 500,000 psi (about 350 kg/mm^2). The ceramics listed here are anywhere from two to two hundred times harder than a "hard" metal. This explains why ceramics such as silicon nitride and zirconium oxide can be used successfully as cutting-tool inserts for machining metals and why aluminum oxide, silicon carbide, and diamond particles are so effective for finishing and polishing metals. It also explains why these same ceramics can provide many times the life of metal parts in applications where wear is a factor.

amazing facts

Diamond is a ceramic and is about 200 times harder than a hardened metal alloy.

This chapter explores the various applications that benefit from the unique hardness of ceramic materials. The chapter is divided into three sections: (1) the cutting of materials, (2) wear resistance and corrosion resistance, and (3) ceramic armor.

THE CUTTING OF MATERIALS

Ceramics are used in many ways for shaping and finishing products, especially metals. Sometimes, the ceramics used to shape an object are in the form of particles that are bonded into a thin wheel for cutting or into a thicker wheel for grinding. Other times, loose ceramic particles (usually referred to as *abrasives*) are bonded onto paper or cloth for sanding or are mixed with a liquid for grinding, polishing, lapping (achieving a very smooth surface finish with superfine ceramic particles), ultrasonic machining, or waterjet cutting. Lapping, for example, is used to create the extremely smooth surfaces needed for making silicon

MATERIAL	HARDNESS, in pounds per square inch (psi)	HARDNESS, in kg/mm^2
Single-crystal diamond	10.0–13.5 million	7000–9500
Polycrystalline diamond	10.0–12.2 million	7000–8600
Cubic boron nitride (CBN)	5.0–6.7 million	3500–4750
Boron carbide (B_4C)	4.5 million	3200
Titanium carbide (TiC)	4.0 million	2800
Silicon carbide (SiC)	3.3–4.1 million	2300–2900
Aluminum oxide (sapphire or polycrystalline)	2.8 million	2000
Tungsten carbide–cobalt "cermet" (94% WC–6% CO)	2.1 million	1500
Zirconium oxide (ZrO_2)	1.6–1.8 million	1100–1300
Silicon dioxide (SiO_2)	0.8–1.1 million	550–750
Borosilicate glass (such as ovenware)	0.75 million	530

Figure 11-1. Hardness of ceramics and ceramic-based materials, as measured by the Indentation test.

chips, computer disks, and compact disks (CDs). Coarser abrasives on a backing are important for the automotive industry, where many pounds of ceramic abrasives are used for every car that's manufactured.

Waterjet Cutting

Waterjet cutting is a relatively new method for rapidly and precisely cutting metals, plastics, composites, and ceramics. Fine ceramic particles are mixed with water and blasted under about 60,000 psi pressure through a small hole down the center of a ceramic tube (nozzle). The water and ceramic particles exit the nozzle as a very fine stream traveling at very high speed. This fine stream of water and abrasive can cut through just about any material quickly and cleanly. The ceramic nozzle must have exceptional wear resistance. A special tungsten carbide ceramic has been developed for waterjet cutting nozzles. Waterjet cutting can slice through several inches of steel and has even been used to bore tunnels through solid rock.

> **IMPORTANT USES OF CERAMIC ABRASIVES**
>
> **Auto-body finishing**
>
> **Polishing of silicon chips, computer disks, and compact disks (CDs)**
>
> **Sandblasting**
>
> **Waterjet cutting**

Ceramic Cutting Tool Inserts

Most metals machining involves cutting with a lathe, milling machine, or other machine tool. A study in the early 1970s estimated that there were 2,692,000 metal-cutting machines in the United States, with an annual operating cost of about $64 billion. Those numbers are probably small compared to worldwide machines and costs today. Key factors identified in the study that dictated the number of machines and the operating costs were the life of the cutting-tool insert and the speed of cutting. Ceramics have dramatically affected both of these factors.

Cutting tools in the early 1900s were made from high grades of tool steel. They could cut only up to about 100 surface feet per minute (sfpm). Sfpm describes the length of surface of the material being machined that is cut by the tool in one minute. The higher the number, the faster the metal is being machined. The development of tungsten carbide-cobalt(WC–Co) cermets, also called cemented carbides (composites consisting mostly of tungsten carbide ceramic cemented together with about 5 or 6 percent cobalt metal), increased the cutting speed to about 400 sfpm for some alloys. During the mid-1950s, alumina-based ceramic cutting-tool inserts were invented that could cut metals at speeds up to 2000 sfpm (610 meters per minute). An example was a cast-iron sleeve 6 inches in diameter that was machined on a lathe, with the alumina tool insert shaving off a ribbon of metal nearly a quarter of an inch wide and 0.016 inch thick. That's equal to removing a block of cast iron 2 by 4 by 11 inches every minute, a remarkable advancement over steel tools and cermet tools. The alumina cutting tools, however, were brittle and not very tough and weren't as reliable as desired by industrial users. The tungsten carbide–cobalt cermets continued to be the dominant cutting-tool material.

A new, higher-strength alumina and a composite of alumina mixed with titanium carbide ceramic particles were introduced in the 1960s, followed by high toughness silicon nitride, transformation-toughened alumina, and alumina reinforced with silicon carbide whiskers, in the late 1970s and early 1980s. These materials increased the reliability of ceramic tools and allowed machining of a much larger range of metal alloys at higher speed than was possible with the cemented carbide tools or the earlier alumina. With these new ceramic tools, the time to machine a metal part could be reduced by 75 to 90 percent, because the ceramic tools could cut 5 to 10 times more rapidly than previous tools. As a further benefit, the ceramic tools lasted longer and could cut more metal parts per tool insert.

Ceramics also have been effective as coatings on cemented carbide tools. The rate of tool wear for the cemented carbide results from a combination of abrasion and corrosion (chemical attack due to the high temperature during cutting). A thin coating—about two ten-thousandths of an inch (5 micrometers) thick—of alumina, titanium carbide, or titanium nitride on the surface of the carbide tool decreases the rate of corrosion and increases tool life by up to five times.

Diamond and Other Superhard Abrasives

Looking back at Figure 11-1, you can see that many ceramic materials are at least several times harder than zirconia, alumina, silicon nitride, and silicon carbide. Including single-crystal diamond, polycrystalline diamond, and cubic boron nitride (CBN), they are referred to as *superhard abrasives.* These superhard abrasives have been developed in parallel with the other ceramics during the last half of the 1900s and are also important for the cutting and machining of metals. Diamond has particularly captivated peoples' imaginations, because of its incredible hardness and also because of its beauty as a gemstone. A whole book could be written about the efforts of alchemists and scientists to duplicate nature and make diamond.

Diamond has been known for centuries to be a special material. The Roman author Pliny referred to diamond in AD 78 declaring, "Indeed its hardness is beyond all expression...." Often thought of as a symbol of eternity, faceted diamond gemstones are highly valued, especially for wedding rings. But diamond also has become an industrial workhorse for machining metals and ceramics and for drilling holes in the Earth to search for oil and metal-containing ores. These industrial needs, added to the age-old dream of creating precious gemstones, finally led to the first successful synthesis of diamond.

Efforts to synthesize diamond date back to about 1797, when Smithson Tennant determined that diamond consists of pure carbon. Based on the properties of diamond and the way it was found in the Earth, scientists believed that a combination of high temperature and pressure would be required to synthesize diamond in the laboratory, but no one was successful.

One researcher claimed to have produced diamonds in about 1880. He sealed a mixture of paraffin wax, bone oil, and lithium metal in an iron tube, which he then heated until it became red-hot. The paraffin and oil in the tube decomposed into carbon and hydrogen, which modern scientists estimate caused pressures of around 100,000 psi and temperature of around 2732°F (1500°C) inside the iron tube. Most of his attempts resulted in explosions. A few experiments, however, managed to survive several hours at the high temperature and pressure. After the metal cooled, it was dissolved in acid. A few tiny, shiny crystal fragments visible in the debris turned out to be diamonds. No other researchers could reproduce his results, though, and the diamonds were later identified as natural, rather than synthetic. Did he fake his results by putting diamonds into his recipe, or did the diamonds get in as an accident? We'll probably never know, but the story does illustrate the intensity of the efforts to make diamonds in the late 1800s.

Despite many other attempts, no one succeeded in making diamonds until Howard Tracy Hall, at General Electric Company, in 1954. He designed an apparatus that could reach the enormous pressure of about 3 million psi and a temperature of 9032°F (5000°C). Hall heated and compressed a mixture of graphite and iron sulfide in his apparatus and produced small diamond crystals. By 1957, his process for synthesizing diamonds was refined to the point that commercial production began. One important innovation was the addition of nickel metal. With the nickel present, diamond crystals could be produced from graphite in about five minutes. Scientists later learned that any source of carbon worked. In fact, they made diamonds from both peanuts and sugar. The age-old quest of the alchemists had finally been realized.

Diamond is incredibly hard but not very tough. Single crystals can be broken easily by an impact, such as striking them with a hammer or jamming a cutting tool too vigorously against the material being cut. A naturally occurring polycrystalline diamond called *carbonado* is much tougher. Once General Electric had learned how to produce single crystals of diamond, they also figured out how to produce synthetic carbonado, which they named *Compax*. An alternate synthetic polycrystalline diamond called *Megadiamond* is produced by Megadiamond Industries in Provo, Utah. Polycrystalline diamond compacts have become important for the machining of non–iron-based metal alloys, polymers, and composites and also as surface coverings for industrial parts that must survive very challenging conditions of wear.

Diamonds produced by squeezing graphite at high pressure and temperature in a small, enclosed chamber are still expensive and can only be made in small sizes. Another dream of scientists has been to deposit diamond from a carbon-containing gas directly onto the surface of another material. William G. Ebersole of the Union Carbide Corporation actually succeeded at this task in 1952, even before high-pressure, high-temperature methods were successful at GE, but the rate of diamond deposition was too slow to be of commercial interest. By the late 1980s, however, scientists had found ways to speed up the rate of deposition by over 100 times, so that a layer a couple thousandths of an inch thick could be achieved in an hour. This still sounds like a long time to build up such a thin layer, but it's fast enough to open many commercial opportunities. Diamond is so hard and durable that only a very thin layer is needed to defeat wear and greatly increase the life of a piece of equipment.

Diamond coatings are now used to extend the life of cutting tools. Some of these coated cutting tools have been able to keep cutting 10 to 20 times longer than the same material without the diamond coating. Engineers also are looking for ways to use the diamond coatings to increase the life of industrial equipment, by reducing wear of critical parts. We are on the threshold of a new revolution in wear-resistant materials. Soon, we'll be able to broadly and inexpensively apply thin coatings of diamond, the hardest material known, to protect a wide variety of other materials.

Figure 11-2. *Silicon carbide ceramic seals, showing the surfaces after rough grinding and final grinding.* (Photo courtesy of Saint-Gobain/Carborundum Structural Ceramics, Niagara Falls, NY.)

WEAR RESISTANCE AND CORROSION RESISTANCE

The cutting of metals is important, but there are many other ways that hard ceramics can be beneficial. This usefulness is due partly to the hardness, but also to the chemical resistance (corrosion resistance), of these same ceramics. The combination of wear resistance and corrosion resistance opens many opportunities in our daily lives, in industry, in cars, and in aircraft. Let's look first at some of the types of parts that can be made of ceramics, and then let's explore the use of these parts and other parts in industries and products that are important to all of us.

Ceramic Seals

Many pieces of equipment, such as water-faucet valves, air-conditioner compressors and water pumps in cars, main shafts in jet engines, sand filters for swimming pools, and assorted pumps and valves, require a seal between sliding or rotating parts to prevent fluids from leaking. The seals for these parts must be able to resist wear from sliding, as well as erosion caused by fluids containing abrasive particles and corrosion from various chemicals. Sometimes, they even must work at high temperature, as do the seals in jet engines and in pumps for molten metals. In addition to being resistant to erosion and corrosion, seals usually have to be exceptionally smooth, to provide easy sliding movement while still preventing leakage. Because ceramics are so hard, rigid, and chemically stable, they make terrific seals.

Figure 11-3. *Variety of silicon carbide seals.* (Samples from Saint-Gobain/Carborundum Structural Ceramics. Photo by D. Richerson.)

Ceramic Valves

Valves are essentially a type of seal. There are many types of valves—rotary, ball and seat, butterfly, gate, and concentric cylinders—and each of these has variations. One type of *rotary valve* is shaped like a ball but has a hole through it, plus a handle that protrudes from the side. When the valve is open, the hole is lined up with the pipe carrying the liquid. To close the valve, the handle is rotated so that the hole is no longer lined up with the pipe but instead faces the sides of the pipe.

A *butterfly valve* is like a door that swings open and closed. It consists of a flattened plate or diaphragm (rather than a ball) that can be swung or turned into or out of the path of the fluid. The butterfly valve is often used to control flow in larger-diameter passages such as air ducts, whereas the rotary valve is used frequently to control flow of a liquid in pipes and hoses. Both of these types of valve can control flow in both directions, at any level between maximum and zero.

A *gate valve* slides in and out of a passage to either block or allow flow. The gate in an irrigation ditch is a gate valve. A gate valve that definitely requires ceramics is the slide-gate valve in a ladle holding molten steel. The slide-gate is opened to pour molten steel into a casting mold, as we discussed in the last chapter. An example of a *cylinder-in-cylinder valve* is the oxygen valve that was illustrated in Chapter 8 for hospital respirator equipment. Another important valve is the type discussed and shown in Chapter 9 for gasoline and diesel engines. This type of valve has a shape that matches the shape and size of a passage or opening and can be moved in and out of the opening. In the case of car engines, one set of these valves opens to let a gasoline–air mixture into the cylinders to be ignited by the spark plugs, and another set opens to let the exhaust gases out of the cylinders.

Figure 11-4. *Silicon nitride check valve balls ranging from 3/4 inch to 1 1/2 inches in diameter.* (Photo courtesy of Ceradyne, Inc., Costa Mesa, CA.)

Pumps

Ceramic seals and valves are used extensively in pumps. Pumps require a valve that allows fluid to pass in one direction only, such as a *ball and seat valve*. A ball and seat valve consists of a spherical ball resting in a tapered cylindrical opening. You can picture this by placing a tennis ball on top of a glass. While the tennis ball is in place, fluid is either held in the glass or prevented from getting into the glass. In the pump, pressure is applied to the fluid in the cylinder under the ball. This pushes the ball away from the cylinder and lets fluid flow past the ball. But as soon as the pump cycle is over and the pressure is released, the ball drops back into contact with the cylinder, which prevents fluid from flowing back into the pump. A ball and seat valve is sometimes called a *check valve* because it

only allows fluid flow in one direction and closes, to check the flow, if the fluid tries to flow backwards.

Pumps use not only ceramic seals and valves for wear resistance and chemical resistance but also ceramic bearings, plungers, rotors, and liners. The materials for these parts must survive severe conditions. For example,

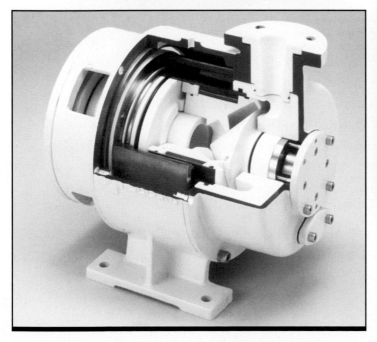

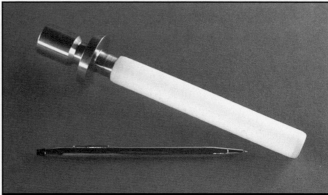

Figure 11-5.(top left) *Silicon carbide pump parts.* (Samples supplied by Saint-Gobain/Carborundum Structural Ceramics, Niagara Falls, NY. Photo by D. Richerson.)

Figure 11-6.(left) *Chemicals-industry pump, completely lined with ceramics.* (Photo courtesy of Kyocera.)

Figure 11-7.(above) *Alumina pump plunger bonded to metal.* (Sample supplied by Coors Ceramics, Golden, CO. Photo by D. Richerson.)

pumps in the mining industry must handle fluids containing large amounts of solid particles that are continually trying to gouge and scratch the pump parts. Pumps in the chemicals and petroleum industries are subjected to corrosive chemicals, often at temperatures from 100° to 800°F. There are even pumps in the aluminum industry that pump molten aluminum and pumps in the papermaking industry that handle hot caustic soda (sodium hydroxide) solutions. A pump with ceramic parts often will last at least 10 times longer than one made of other materials.

Ceramics in the Mining Industry

The mining industry is a high-wear industry. Rock is continually being crushed, ground, conveyed, leached, and screened in an effort to separate the metals or other material of value from the waste rock. All of these operations are either abrasive or corrosive. Ceramics would be great for their wear resistance and corrosion resistance, but often they can't be used because they can't withstand the impact from big chunks of rock or can't be fabricated into the enormous pieces of equipment needed. For pieces of equipment that aren't exposed to too high an impact, ceramics are used as linings. Tiles of ceramic, usually alumina, about the size of bathroom floor tiles are attached to metal surfaces with a glue such as epoxy. This technique has been successful for lining ore chutes, conveyors, and tanks containing corrosive chemicals. Ceramics also are used in pumps and valves, in equipment that handles dust, and in equipment that separates fine particles of rock and ore from water (such as centrifugal separators in the mining industry).

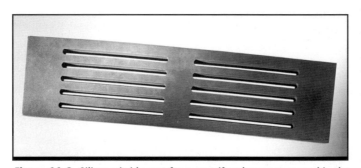

Figure 11-8. Silicon nitride part for a centrifugal separator, used in the mining industry to separate water from chunks and particles of rock. (Sample from Ceradyne, Inc., Costa Mesa, CA. Photo by D. Richerson.)

Ceramics for Power Generation

Much of the electrical power in the United States is generated by burning coal. Crushing and conveying the coal results in wear problems similar to those in the mining industry. One especially challenging area involves transferring coal that has been crushed into a fine powder from the crusher to the large furnace where the coal is burned. This powdered coal typically is conveyed by blowing it with air (pneumatically) through tubes or pipes. Metal pipes in some power plants had been wearing through in about a week. Silicon nitride ceramic linings recently were installed, and no replacement has been required for over a year. This situation points out a dilemma for the ceramic manufacturers: Sometimes ceramics last too long! Like in the Maytag man advertisements, there is little or no return business.

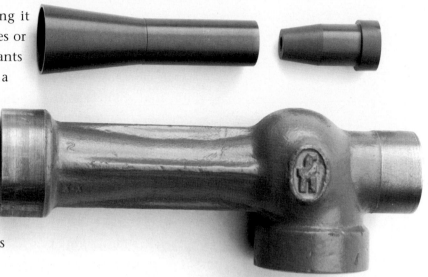

Figure 11-9. Silicon nitride Eductor liners and metal holder for the pneumatic conveying ot powdered coal. *(Photo courtesy of Ceradyne, Inc., Costa Mesa, CA.)*

Ceramics in Papermaking

So far we've only explored a few of the uses of ceramics in selected industries. Most industries use ceramics in almost every step of producing a product from a raw material. Let's pick one industry and take a more in-depth look. How about papermaking? When you read your newspaper or use a paper towel, do you suspect that ceramics played any role in their manufacture? You'll be surprised at how large a role ceramics do play and also at the complexity of the papermaking process.

The logistics of papermaking are impressive. On a typical day, a paper mill produces around 1500 tons of paper, consumes enough electrical power to provide all the needs of a city of 40,000 people, and uses about 65 million gallons of water (roughly equivalent to the amount that flows over Niagara Falls in 90 minutes).* With an operation of this size, shutdown time for maintenance is very expensive. As a result, the paper industry uses ceramics extensively, because of their superior resistance to wear and corrosion.

Figure 11-10 shows a schematic of one of the major papermaking processes, called the *Kraft process*. This complex process can be divided into four stages, or steps: (1) preparation of wood chips in the woodyard, (2) conversion of the wood chips to pulp, (3) conversion of the pulp to paper, and (4) recycling of chemicals, energy, and water. Let's look at the different things that happen in each stage and the way ceramics are involved.

Activities in the Woodyard. In the woodyard, logs are cut to workable lengths, stripped of their bark, and chopped into chips about one inch

---

*Statistics and much of the discussions of ceramic applications for papermaking are based heavily on input from Richard DeWolf, Thielsch Engineering, Inc., Rock Hill, SC.

across. Raw logs (straight off the logging truck) first pass sideways through a series of 6 to 8 foot (1.8 to 2.4 meter) diameter circular saws that cut the logs into shorter lengths. The saw blades are tipped with tungsten carbide–cobalt cermet cutting edges. The logs then are loaded into a debarking drum that rotates, so that the logs rub against each other until their bark comes off. Finally, the stripped logs go through the chipper. The chipper has two main parts. One is a heavy disk that contains anywhere from 4 to 16 knife blades and spins at 1800 rpm. Next to this rotating disk is a series of nonrotating blades, or plates. The chipper can convert a log 5 feet long and 1 foot in diameter into chips in 2 to 3 seconds. Wear and tear on the blades, especially the stationary knife blades is extreme. Metal blades generally need replacement at least once a week. A composite material consisting of titanium carbide ceramic particles reinforcing a metal can survive for 36 days.

The wood chips are carried on a conveyor belt, or blown pneumatically by air through a large pipe, to screens that separate the chips according to size. The chips of the right size are then conveyed to the equipment that converts them to pulp. Aluminum oxide tiles and blocks are used as wear-resistant surfaces on screens, as liners in pneumatic conveyor pipes, and as liners in equipment that removes sawdust.

Pulp Preparation.   The next stage of the Kraft process is to convert the wood chips into *pulp*. Pulp consists of individual wood fibers mixed with water. To obtain the fibers, the glue (*lignin*) that holds the fibers together in the wood must be removed. This job is done in a large pressure cooker, called a *digester,* that's typically 50 to 250 feet high and operated at 100 to 180 psi pressure. Chemicals such as sodium sulfide and sodium hydroxide are added to help break down the glue. The chips, water, and chemicals are heated with steam. To minimize corrosion problems, some digesters are lined with ceramic brick.

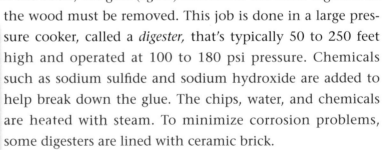

Ceramic–metal composite chipper blades can reduce a 5 foot long log to chips in 2 to 3 seconds and last 7 times longer than metal blades.

At the conclusion of digestion, which takes about one to two hours, a valve is opened to let the pressurized contents of the digester spray violently into a tank called a *blow tank.* The rapid release of pressure as the chips leave the digester causes them literally to explode into their individual fibers, which mix with the water and chemicals to form raw pulp. This raw pulp is washed, in a series of steps, to separate the fibers from the chemicals and lignin. Because the chemicals are corrosive, the large vats where the pulp is washed and the large tanks where the pulp is stored are lined with ceramics. Pumps, valves, and seals are necessary between each

Figure 11-10. *Schematic of the Kraft papermaking process.* *(Printed with the permission of the Westvaco Corporation, New York, NY.)*

washing cycle. These devices also contain ceramic parts to resist wear and chemical attack.

After it's washed, the pulp is brown in color, suitable for making cardboard and shopping bags but not for making white paper products. For these white products, the pulp must be bleached. Bleaching involves treating the pulp with more chemicals, just like when you add a fabric bleach to your washing machine. One bleach chemical and one cycle aren't enough to get the pulp white. Four or five cycles with strong bleaches such as chlorine, sodium hypochlorite, chlorine dioxide, hydrogen peroxide, and oxygen are necessary. Some of the bleaching tanks require ceramic tile linings to protect against these strong chemicals. Additional washing steps, using water, also are required to remove the bleaches. More pumps and valves with ceramic parts are necessary to transfer the pulp from tank to tank for these bleaching and washing steps.

The bleached pulp moves to large storage tanks that can hold 500 to 1500 tons of *stock,* a mixture of about 90 percent water and 10 percent wood fibers. These enormous tanks are made of concrete (a ceramic) lined on both the inside and outside with ceramic tiles. The stock is diluted to about 3 percent wood fibers and passes through beaters, or *refiners,* that cut the wood fibers even smaller, to produce a smooth, uniform pulp slurry. This pulp slurry is then further diluted with water, to about one-half a percent of wood fibers, and finally is ready to feed into the paper machine.

The Paper Machine. The paper machine is amazingly complex, a remarkable feat of engineering. It typically takes a 20 foot (6 meter) wide flow of the pulp slurry containing only one-half a percent of fiber, removes all of the water, and compacts the wood fibers into smooth, dry paper in about 30 seconds. The slurry enters the paper machine through a long, narrow slit that measures just the right thickness for the pulp needed to make the particular grade of paper. This pulp flows onto a moving sheet of clothlike fabric called the *wire.* The wire is made of high-strength synthetic fibers, such as nylon, woven into an open mesh somewhat like window or door screen. The wire passes over ceramic slats, called *foils,* and ceramic plates, called *suction box covers,* that contain open slots or holes. Water is sucked between the foils and through the suction box cover holes, while the wood fibers continue moving through the paper machine on the wire. Ceramics are important for these steps because of their wear resistance and smoothness. The fabric

CERAMICS IN PAPERMAKING

Cutter edges

Wear-resistant liners for chutes, screens, pipes, and conveyors

Corrosion-resistant liners for vats, troughs, sewers, chemical recyclers, and bleach tanks

Pump components

Seals and valves

Cyclone cleaner liners

High-temperature refractory linings

Coatings

Foils and suction box covers

Sensors

Electronics

screen is traveling at nearly 80 miles per hour and could be damaged by a rough surface or sharp edges. Suction box covers are made from aluminum oxide. Alumina, transformation-toughened alumina or zirconia, and silicon nitride have all been used for foils.

After passing over the foils and suction boxes, the partially dewatered pulp (still containing about 80 percent water) moves off the wire onto a sturdier fabric carrier called a *felt* and passes through a series of large rollers (called *press rolls*). These press rolls compact the loosely entwined wood fibers into a solid sheet and squeeze out more water. The primary press roll traditionally has been made of granite rock, a ceramic made by Mother Nature, and quarried out of a mountainside as a single piece. The largest of these rolls is over 20 feet long and about 5 feet in diameter, with a steel shaft through its center. Because these rolls have occasionally failed violently, destroying surrounding pieces of the paper machine and sometimes injuring workers, ceramic coatings on steel are being developed as a safer alternative to granite.

Figure 11-11.(top) *View of ceramic foils and suction box covers installed on a paper machine.* (Photo courtesy of Coors/Wilbanks, a Coors Company, Hillsboro, OR.)

Figure 11-12.(bottom) *Aluminum oxide ceramic suction box covers being assembled for shipping and installation.* (Photo courtesy of Coors/Wilbanks, a Coors Company, Hillsboro, OR.)

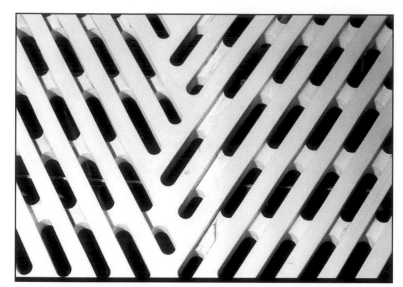

Figure 11-13. *Close-up view of ceramic suction box covers.* (Photo courtesy of Coors/Wilbanks, a Coors Company, Hillsboro, OR)

Figure 11-14. *Silicon nitride segments ready to be machined and assembled into papermaking foils.* (Photo courtesy of Ceradyne, Inc., Costa Mesa, CA.)

Once the pulp has been compacted by the press rolls, it looks more like paper but still contains a lot of water. The paper now goes through another series of rolls that are heated to drive off the remaining water. To be completely dried, some grades of paper must pass between as many as 120 sets of rolls. All of these heavy steel rolls are expensive to buy and to operate. Engineers currently are testing ceramic-coated rolls that can substantially reduce the number of drying rolls and reduce costs.

Some grades of paper can be cut to the right size after drying and packaged for shipping to stores. The better grades, such as writing and typing paper, still have to face another important step; they must be coated with fine particles of clay or other ceramic to give the paper extra smoothness and brightness. The ceramic particles are applied to the surface of the paper as a water-based slurry. The water is quickly evaporated with heaters, some that are ceramic. Finally, the paper is cut to the desired width with ceramic *slitters* and wound onto a drum or spool to be shipped and packaged.

Recycling of Chemicals, Water, and Energy. The fourth stage of papermaking is recycling the chemicals, water, and energy. Bark, wood scrap, and lignin are burned to heat boilers to make steam and also to generate electricity. Chemical vapors released by this burning are condensed (cooled and removed as a liquid) to produce turpentine and other chemicals. Sodium hydroxide is recovered from unburned slag (called *smelt*) using calcium oxide ("quick lime"), which is also then recovered in a lime kiln. Ceramics are used in a number of these operations, especially in the lime kiln, which reaches temperatures above 1800°F. A typical lime kiln is 8 to

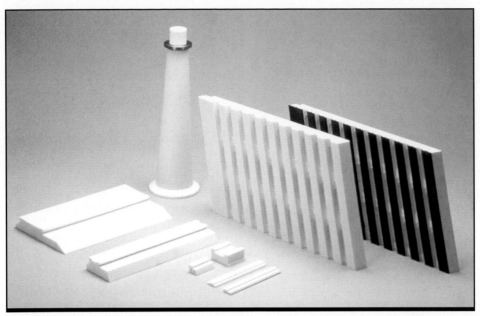

Figure 11-15. *Variety of papermaking ceramics.* (Photo courtesy of Kyocera.)

15 inches in diameter and 75 to 350 feet long and completely lined with wear-resistant, heat-resistant ceramic refractories.

Are you surprised at the complexity of papermaking and the number of ways that ceramics are used? The paper industry depends heavily on ceramics, as do most other industries, yet the ceramics are hidden, so that most people aren't aware of their important role.

Ceramics in the Textile Industry

Ceramics are also vital to the textile industry, especially for the manufacture of thread and in automated weaving machines. Automated machines move thread at high speed, under tension, from spindle to spindle, so that the thread slides over many guiding surfaces. There are thousands of these *thread guides* in a textile plant. Thread guides must be very smooth, so that they don't fray or damage the thread. They also must be very wear resistant, because rapidly sliding thread can cut a notch through most metals and

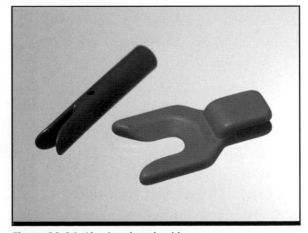

Figure 11-16. *Alumina thread guides.* (Samples from Diamonite Ceramics, about 1985. Photo by D. Richerson.)

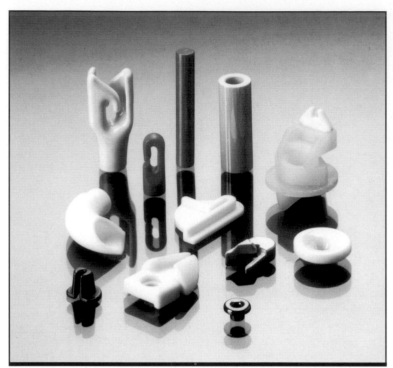

Figure 11-17. Glass-coated and other thread guides. (Photo courtesy of Rauschert Industries, Inc., Madisonville, TN.)

plastics very quickly. Alumina-based ceramics, and sometimes porcelain or stoneware ceramics with glass coatings, are used for thread guides.

Ceramics for Oil-Well Drilling and Pumping

Exploration for oil is a high-technology endeavor and uses ceramics in diverse ways, ranging from drilling "mud" and cement (to help prevent the drill hole from collapsing) to components in a down-hole telemetry system for guiding the drill bit during drilling. Drilling through solid rock is accomplished with diamond crystals or carbonado particles embedded in the surface of a steel drill bit. The bit is mounted on the end of a string of steel rods, which are rotated with a high-horsepower engine to allow the bit to grind its way through the rock. Wells over 5000 feet deep have been drilled. Sophisticated telemetry systems are mounted next to the drill bit as it cuts through the rock. The telemetry system consists of sensors and a transmitter to continually send information to the surface, so that the drillers know the conditions at the bottom of the drill hole. Ceramics are required for several components in the telemetry system, some of which are shown in Figure 11-18.

Once a well has been completed, surface pumps are not always able to pull the oil to the surface. A special pump, referred to as a *sucker rod,* is needed at the bottom of the well to give the oil a boost. This pump contains a ball and seat type of valve. During each pump cycle, the ball

Figure 11-18. Silicon nitride ceramic parts for down-hole oil applications: sucker-rod balls, impulse rotor for a telemetry system, and an odometer wheel for down-hole measurements. (Photo courtesy of Ceradyne, Inc., Costa Mesa, CA.)

is forced upward, away from the seat, to allow oil, gas, hydrogen sulfide, sulfur oxides, sand, and whatever else is mixed with the oil to pass through, and then is pushed back against the seat under all of the weight of the oil in the pipe between the sucker rod and the surface. The ball and seat must withstand severe stress and impact, as well as abrasion from the sand particles and attack from the corrosive fluid at temperatures up to about 200°F. Until the 1980s, no ceramics could survive these severe conditions, so steel balls with tungsten carbide–cobalt seats were used. Today, the balls are made of high-strength silicon nitride

Figure 11-19. Silicon nitride wire-drawing tooling. *(Photo courtesy of Ceradyne, Inc., Costa Mesa, CA.)*

or transformation-toughened zirconia and provide five to ten times the life of steel balls.

amazing facts

Aluminum beverage cans are produced with hard ceramic tooling.

Other Wear-Resistance Applications for Ceramics

Hard ceramics are used in countless other ways, such as for tools that produce metal wire and form aluminum cans and for special coatings on magnetic recording disks (computer disks and CDs). The wire-drawing tooling converts metal rods into various sizes of wire through a combination of stress, pressure, and temperature. Alumina, silicon nitride, and transformation-toughened ceramics all have been used successfully for wire-drawing tooling.

Have you ever wondered how aluminum beverage cans are made, especially the pop-top openers? They're created using a sequence of ceramic tools. The tools stamp, bend, extrude, fold, and crease, and do all of the other operations necessary to shape and seal the aluminum. These tools are exposed to high stress and severe wear and must have hard, smooth surfaces that cause little friction against the thin aluminum metal. Silicon nitride, transformation-toughened ceramics, and alumina reinforced with silicon carbide whiskers all have been used successfully to produce aluminum cans.

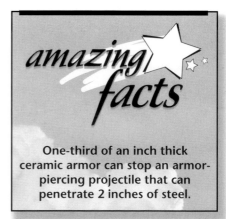

amazing facts

Hard ceramics are required in all CDs, computer disks, and read/write and recording heads.

Computer disks and CDs are metal disks coated with layers of a hard ceramic and a special information-storing magnetic material. The hard ceramic used on a disk must be very smooth, rigid, and precise in size and shape. Silica, zirconia, and diamond-like carbon all have been used successfully as the ceramic coating on disks. Researchers are working right now to make entire disks from ceramics. The head, which reads information off a disk or transfers information onto the disk, also requires a hard ceramic. A ceramic composite of titanium carbide particles in aluminum oxide has been the standard material for years.

CERAMIC ARMOR

So far we've explored how the hardness of ceramics makes them ideal for resistance to wear, which probably hasn't been a big surprise to you. Another use which takes advantage of the hardness in a much more surprising way, though, is bulletproof armor. Amazingly, a flat plate (only about one-third of an inch thick) of some ceramic

amazing facts

One-third of an inch thick ceramic armor can stop an armor-piercing projectile that can penetrate 2 inches of steel.

materials can actually stop a bullet able to penetrate more than 2 inches of steel; yet, these same ceramic materials will shatter if dropped on the floor. How is this possible? The secret is the hardness of the ceramic, plus the way it's held in place. The ceramic plate is bonded onto a plate of fiberglass with a glue similar to rubber cement. When a bullet strikes the ceramic, the ceramic is hard enough to shatter the bullet into little pieces. The ceramic near where the bullet hits also shatters, which helps to absorb some of the energy of the bullet. The fiberglass catches all of the broken pieces of the bullet and ceramic, just like a baseball mit catches a ball. The person wearing the armor may get pretty bruised but at least can walk away from what otherwise would probably have been a fatal experience.

Ceramic armor was developed during the Vietnam War to protect soldiers and helicopter pilots from small-arms fire. The ceramic that was developed then was boron carbide, which is nearly twice as hard as alumina and has less than one-third the weight of steel. Boron carbide is still an important armor material for soldiers, police officers, and helicopters. Alumina and a composite of alumina plus aluminum also work as armor but are heavier than boron carbide and need to be thicker because they aren't as effective as boron carbide at stopping bullets. These materials are used on land-based military vehicles.

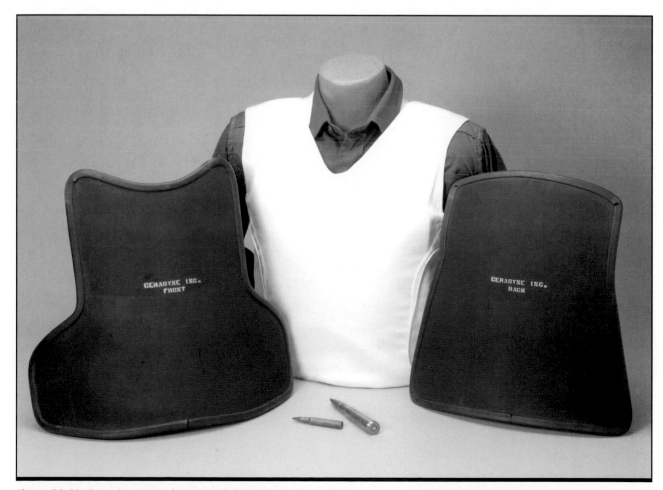

Figure 11-20. *Ceramic personnel armor and the projectiles it can defeat.* (Photo courtesy of Ceradyne, Inc., Costa Mesa, CA.)

OVERVIEW: CERAMICS STOP WEAR AND BULLETS

Perhaps you now understand why ceramics can be called the hardest materials in the universe. Their combination of hardness, strength, light weight, smoothness, and chemical resistance help keep our manufacturing industries, cars and airplanes, and electrical power generation plants running smoothly. Despite their many successful uses, though, ceramics often are not used in other applications either because they're either too expensive or out of fear that they might break and cause damage to other parts. The new ceramics like silicon nitride and transformation-toughened zirconia are proving that they can be reliable and not break, and their cost is coming down to the point that they will be used much more broadly in years to come. The development that really stimulates the imagination of engineers, however, is diamond coatings. Don't be surprised to see diamond coatings showing up on all kind of products, to increase product life and performance; they're already being used on razor blades to give a cleaner, smoother shave—so the opportunities are unlimited.

Energy and Pollution Control

*P*opulation growth and industrialization have led to enormous growth in our world energy requirements. To meet that demand, we've often made compromises that have wasted natural resources and produced serious pollution problems. Ceramics are now playing an increasingly important role in enabling alternate, cleaner sources of energy, improving the efficiency of our present power generation methods, and reducing the amount of pollutants we release into our air and environment. This chapter addresses these topics, as well as the role of ceramics in energy conservation.

THE ROLE OF CERAMICS IN ENERGY CONSERVATION

> **PRODUCTS AND USES**
>
> Fiberglass insulation
> Double-paned, tinted glass
> Efficient lighting
> Efficient, low-pollution cars
> Gas-appliance igniters
> Industrial heat exchangers
> Improved heat engines
> Clean-coal energy systems
> Nuclear fuels and controls
> Nuclear waste encapsulation
> High-efficiency fuel cells
> Power from waste incineration
> Motor-vehicle emissions control
> Industrial emissions control
> Water and sewage treatment
> Oil-spill cleanup

We've been incredibly wasteful of our natural resources, such as coal and oil, especially during the last century and since the invention of the automobile. Coal and oil are nonrenewable resources; they don't regrow or replenish themselves. When they're gone, we'll have to look elsewhere for the energy to keep our cars running, to heat our homes and offices in the winter, and to provide the electricity for industry and other needs. We have made important strides in energy conservation, though, particularly since about 1970. The historical event that really caught our attention, early in the 1970s, was the oil embargo by the Organization of Petroleum Exporting Countries (OPEC). OPEC limited the amount of oil it would export. For the first time, industrialized countries such as the United States suffered an oil shortage; and for the first time, the general public felt the shortage. Products made from petroleum, such as home heating oil and gasoline, jumped in price and were much more difficult to get than ever before. We even had to wait in long lines to put gasoline in our cars, as an example.

The OPEC crisis stimulated the development of a national energy policy in the United States, directed towards energy conservation, alternate energy sources, and increased use of renewable sources of energy such as wind, water, geothermal resources (heated water from volcanic activity), biomass (energy from plant matter), and sunlight. Ceramics and glass have played an enormous role in these programs and in all aspects of energy conservation.

Fiberglass Insulation

Although the OPEC crisis led to many new programs, some earlier programs had already made a big difference in energy conservation. One of the most important was the invention of fiberglass thermal insulation for use in the walls and attics of homes and buildings, to keep the heat inside during the winter and outside during the summer. The development emerged from a partnership between Owens-Illinois and Corning Glass Works from 1931 to 1938. They invented a process for producing tiny, hair-thin strands, or *filaments,* of glass—somewhat like cotton candy or steel wool—that intertwined loosely into bundles that trapped a lot of air. Such "dead air space" is a good barrier to heat, just like the air trapped in the pores of the Space Shuttle tiles. You've probably seen this insulating material in builders' supply stores, in your own home, or at construction sites. It's usually bright pink and sold in large rolls.

Fiberglass insulation is produced by melting round marbles of refined glass in an electric furnace. The melted glass is allowed to flow into a heated chamber with hundreds of small holes in the bottom. The molten glass flows through these holes into another chamber below. As the glass flows in thin rivulets down into this lower chamber, it's blasted with jets of high-pressure steam. The rapidly moving steam stretches each molten rivulet of glass into a thin fiber and cools it quickly into solid glass. These many strands form the loose bundle of glass fibers that works so effectively for insulating our homes. Each marble melted in the glass furnace is about five-eighths of an inch in diameter and produces about 97 miles of fiber.

amazing facts

Fiberglass insulation was introduced in 1938 and has saved over 25,000,000,000,000,000 Btu of heat energy from being lost through the walls and ceilings of houses, enough to provide all of the energy needs of North America for two years.

How much energy has been conserved by insulating houses with fiberglass since its invention in 1938? The answer wasn't readily available when I asked that question. After numerous telephone calls, I finally found someone willing to make a rough estimate, Dr. Arun Vohra of the Office of Buildings Technology at the U.S. Department of Energy. Using a few assumptions, and data from a 1993 government report on household energy consumption and expenditures, he estimated that roughly 25 quadrillion (25,000,000,000,000,000) Btu of heat energy has been conserved, saving an average of more than $2 billion per year. That's enough energy to supply all the energy needs of the United States for a year! Dr. Vohra's estimate was conservative: He assumed that only 50 percent of homes have fiberglass insulation, and he didn't include apartments and buildings. The total energy conserved by using fiberglass insulation is probably even higher than his rough estimate.

Double-Paned Glass and Reflective Coatings

Fibrous insulation is great for walls and attics, but it is opaque and can't be used to insulate windows. To decrease heat loss through windows, manufacturers developed double-paned glass storm windows that provide a dead air space between two panes of glass. Further improvements have been achieved by coating window glass with a very thin layer of a reflective metal that allows most of the light to pass through but reflects much of the heat. This coating especially helps keep a house or building cool during the summer. Many high-rise office buildings are now entirely covered with glass with reflective coatings. As shown in Figure 12-1, these buildings

Figure 12-1. *Building with double-paned glass and reflective coating to conserve energy.* *(Photo by D. Richerson.)*

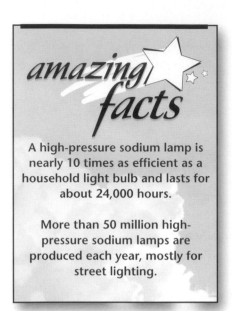
with reflective glass are architecturally beautiful, as well as energy efficient.

Efficient Lighting

You may be a frequent traveler, like me, who spends much time flying into cities at night. If so, you've probably noticed the astounding number of lights visible from the air, especially in and around large cities. Those lights consume an enormous amount of electrical energy each night. Fortunately, major innovations made possible by ceramics have increased the efficiency of lighting.

The first widely commercialized lighting was Thomas Edison's *incandescent light,* the basic type of light we have in our reading lamps at home. Modern incandescent lights consist of a tungsten metal filament curled up like a tiny spring and mounted inside an airtight glass bulb. You can look inside a nonfrosted light bulb and easily see the filament and how it is mounted. Tungsten is one of those materials we talked about in Chapter 6 that conducts electricity a little but has high enough electrical resistance that most of the electricity is converted to heat. When the electrical switch is turned on, the thin tungsten filament in an incandescent light heats up rapidly, until it's glowing brightly enough to light a room.

Incandescent lighting, however, isn't very efficient, and the tungsten filament often burns out. The fluorescent light described in Chapter 4 was a big improvement, because it was more efficient and lasted much longer

Figure 12-2. *HPS lamps lighting a highway (left) and a street in Seattle (right). (Photo courtesy of OSRAM SYLVANIA Products, Inc., Beverly, MA.)*

Figure 12-3. *High-pressure sodium vapor lamps, showing the alumina arc tube inside the glass bulb.* (Photo courtesy of OSRAM SYLVANIA Products, Inc., Beverly, MA.)

than earlier lights. An even better light source is the high-pressure sodium (HPS) vapor lamp, which was invented in the 1960s. These are the lights that give off golden or yellowish light and are used for street lighting. An HPS lamp produces 140 lumens (the unit by which light output is measured) of light per watt of electricity, compared to only 15 lumens for an incandescent light. An HPS lamp lasts about 24,000 hours under all weather conditions—that's about six and one-half years if the light is operated for 10 hours each day. Incandescent lights are lucky to last 1000 hours (except for the one over our basement stairway, which requires an acrobat to change and seems to burn out every 200 hours).

More than 50 million HPS lamps are now produced worldwide each year. The key component in an HPS lamp is a thin-walled *arc tube,* which is the container for high-pressure gases (vapors) of sodium and mercury. A typical arc tube is about five-sixteenths of an inch in diameter and about four inches long and fabricated with the same high-production rate type of equipment used for making spark plug insulators. The arc tube is mounted in a larger, sealed glass bulb, as shown in Figure 12-3.

When an HPS lamp is turned on, electricity arcs across the tube through the vapors, giving off light but also heating the tube to temperatures around 2200°F. Developing a material that was transparent yet could withstand the high temperature, pressure, and corrosiveness of the vapors was a big challenge. Glass didn't work. The solution was alumina.

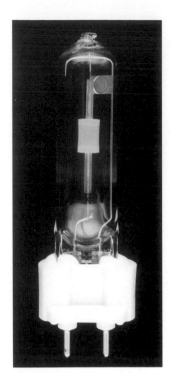

Figure 12-4. *Seventy-watt metal halide lamp with a ceramic arc tube.* (Photo courtesy of OSRAM SYLVANIA Products, Inc., Beverly, MA.)

Scientists learned during the early 1950s how to make polycrystalline alumina that was about 70 percent transparent, which turned out to be good enough for sodium vapor lamps. This same translucent alumina was adapted in the 1980s to make the orthodontic brackets that we learned about in Chapter 8.

Colors don't look the same under HPS lights as they do in daylight. Another type of high-intensity lighting produces white light, so that colors appear more natural. This is a halogen lamp, which has an arc tube filled with a *metal halide* vapor rather than sodium vapor. Although metal halide lights produce whiter light, they're more expensive than HPS lamps and don't last as long. The metal halide vapor in current halogen lamps is sealed into a fused silica (silicon dioxide glass) arc tube. Researchers are exploring alumina arc tubes and other technologies in an effort to make metal halide lamps that are more efficient and last longer.

Automobiles with Increased Gas Mileage

Most of the petroleum we import is processed in large refineries to produce gasoline for our motor vehicles. This use of petroleum is a big contributor to our large trade deficit (the difference between the value of goods we import and the value of goods we export), which adversely affects our national economy. Because of these factors, motor vehicles were a major target for conservation efforts following the 1970s oil crisis. The average gasoline mileage for automobiles has increased dramatically during the past 25 years. Part of this improvement has been achieved by producing

Figure 12-5. *George Washington Bridge lighted with metal halide lamps and contrasting HPS lighting in the background.* (Photo courtesy of OSRAM SYLVANIA Products, Inc., Beverly, MA.)

smaller cars that don't guzzle as much gasoline. The introduction of oxygen sensors and engine electronics control systems, which you learned about in Chapter 9, has also caused a significant improvement in mileage. Ceramics have also helped reduce the weight of cars, primarily by enabling us to replace heavy, iron-based metal alloys with lightweight, ceramic-reinforced composites.

Ceramics also are playing a critical role in the development of our next generation of automobiles, which will have gas mileage of around 80 miles per gallon. These automobiles will be lighter weight, will have improved-efficiency engines, and probably will have hybrid propulsion systems, consisting of a heat engine (gasoline, diesel, turbine, or Stirling) paired with an electric engine.

Ceramic Igniters for Natural Gas Appliances

Natural gas, which is the chemical compound *methane* (made up of a carbon atom surrounded by four hydrogen atoms), is another important energy source that comes out of oil wells. Rather than a liquid like petroleum, though, methane is a gas, like oxygen or nitrogen. Many home furnaces, stoves, water heaters, and clothes dryers burn natural gas to provide their heat. Up until the early 1970s, all natural gas appliances required a pilot light, a small flame that burned continuously so that you didn't have to re-light the appliance each time you used it. A study conducted in

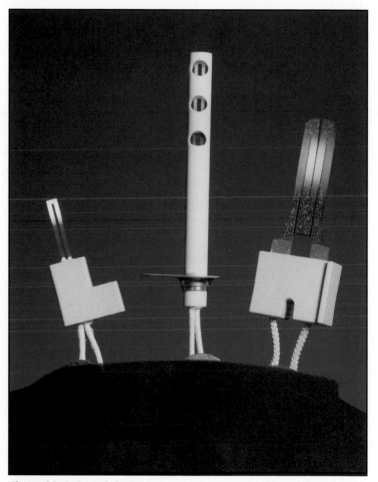

Figure 12-6. Ceramic igniters. *(Photo courtesy of Saint-Gobain/Norton Igniters, Milford, NH.)*

California in the 1960s estimated that pilot lights wasted about 35 to 40 percent of the natural gas that came into the home, which equaled about 6 million cubic feet of natural gas wasted each day, just in the United States. California passed an energy-conservation law that pressured industry into coming up with an alternate technology to replace the pilot light.

The answer, developed in the late 1960s, was the ceramic *igniter*. A variety of ceramic igniters are shown in Figure 12-6. They are made from a silicon carbide ceramic composition that is a semiconductor. The igniter is another of those materials, like the tungsten filament in the incandescent light, that gets really hot as electricity tries to get through it. When you turn on an appliance with an igniter, electricity enters the igniter and causes it to heat, within a few seconds, to a temperature high enough to ignite natural gas. A timer opens a valve to let the natural gas into the appliance about seven seconds after you turn it on. The igniter lights the gas and then shuts off. The pilot light is completely eliminated, saving a tremendous amount of natural gas.

A few ceramic igniters became available for testing in gas clothes dryers in 1968. By 1980, over 750,000 igniters were being manufactured per year. Igniters for gas ranges became available commercially in about 1974 and reached a market level of about 1,700,000 per year by 1980.

Ceramics for Industrial Energy Conservation

Heat Exchangers. As you learned in the previous chapter, most industrial processes have at least one step that involves high temperature. The temperature is achieved by either burning a fuel or using electrical heaters. Another technological thrust in the 1970s was to reduce energy consumption in industrial processes by increasing the efficiency of these high-temperature steps. Much of the effort focused on developing heat exchangers to use waste heat (which normally would go up the furnace stack) to preheat the air used for a process. The more the incoming air could be preheated, the less fuel would have to be burned to get to the temperature needed for the process. Based on the high fuel prices at the peak of the oil crisis and the expectation that they would go even higher, industrial companies were anxious to come up with any technology that would minimize fuel costs. They worked closely with the government to develop new technologies such as ceramic heat exchangers. These programs demonstrated that ceramic heat exchangers could save nearly 50 percent of the fuel. Just about the time success was achieved, however, the price of fuel went down. Many industrial companies decided that the cost of installing and maintaining ceramic heat exchangers didn't economically justify the amount of fuel that could be conserved, so they chose not to install them.

Fuel conservation alone is not incentive enough to make industry change old habits: An overall cost benefit must be clearly seen. Ceramic heat exchangers in the 1970s were expensive. Efforts since then have brought down costs to the point that some industries are starting to use

amazing facts

In the 1960s, pilot lights wasted about 6 million cubic feet of natural gas each day in the United States.

Ceramic igniters replace pilot lights and don't waste natural gas.

By 1980, 1,700,000 ceramic igniters were produced per year for gas ranges.

Figure 12-7. *Silicon carbide heat-exchanger tubes with internal fins to increase efficiency. (Photo courtesy of Schunk-INEX Corp., Holland, NY.)*

them, especially the metals heat-treating industry. Most of these heat exchangers are made of tubes of silicon carbide mounted so that the hot exhaust gases released during processing flow around the tubes. Inlet air for the industrial process flows through the tubes and is preheated. Sections of some silicon carbide heat exchanger tubes are shown in Figure 12-7.

Another type of ceramic heat exchanger with potential for some industrial processes has channels running through it, similar to the honeycomb structure of the ceramic catalyst support shown in Chapter 9 for an automotive catalytic converter. Rather than all of the channels running in the same direction, though, the channels in the heat exchanger alternate directions with each row, as shown in Figure 12-8. The hot gases pass through all of the channels going in one direction, while the cold inlet air passes through the channels in the perpendicular direction. This device is referred to as a *crossflow heat exchanger.* Crossflow heat exchangers have great potential for small industrial applications and also show potential for increasing the efficiency of gas turbine engines.

Figure 12-8. *Crossflow heat-exchanger module fabricated from an NZP composition, with air passages highlighted in blue and combustion gas passages highlighted in red. (Photo courtesy of LoTEC, Inc., Salt Lake City, UT.)*

Radiant Burners.   Another industrial area in which ceramics have provided energy conservation is for radiant heating, especially that achieved using the flat-plate, porous surface burners discussed briefly in Chapter 10. These burners produce very low pollution and are used for drying paint, shaping automotive windows, and heating air for a wide range of industrial processes. A recent advance at a company called Alzeta in Santa Clara, CA, involved using ceramics to greatly increase the efficiency of porous surface radiant burners. By placing an open mesh screen made of a silicon carbide-silicon carbide ceramic matrix composite about one-half inch away from the surface of the burner, Alzeta nearly doubled the radiant heat transfer from the burner. This dramatic increase in efficiency is expected to save a lot of fuel and energy once the burners are installed at industrial sites.

Ceramics and Renewable Energy Sources

Energy from Wind and Water. As mentioned earlier, the burning of oil and coal consumes natural resources that required millions of years to form and that cannot be replenished. We need to switch more and more emphasis to energy sources that can be replenished, referred to as *renewable energy sources*. Two renewable energy sources that have been used for many years are windmills and hydroelectric dams, to harvest energy from wind and water, respectively. Ceramics are important in both technologies. Modern windmill blades are fabricated from composites containing ceramic-fiber reinforcement. Dams generally are constructed from concrete. Both methods of harvesting energy also require electrical ceramics, especially electrical insulators. They also require machinery that rotates at high speed, which can benefit from new wear-resistant ceramics, especially the new silicon nitride ceramic bearings.

Energy from the Sun. An important, but underused, renewable energy source is sunlight. There are many ways of tapping the sun's energy. One way is to focus sunlight with mirrors to produce high temperature to heat a fluid to power a turbine or electrical generator. Another method for tapping the sun's energy is to produce electricity directly with *photovoltaics*, which most of us refer to as *solar cells*. Solar cells are made with semiconductor silicon similar to that used for making silicon chips. As you may remember from Chapter 6, the energy gap of semiconductor silicon is just narrow enough that the energy in the light from the sun can cause electrons to jump across the gap, resulting in a flow of electricity. Once the solar cell has been manufactured and installed, it produces electricity whenever the sun shines—at no cost and with no pollution!

Why aren't we using solar cells to produce electricity for our homes, instead of just for calculators, satellites, and experimental cars? They're still expensive; they aren't very efficient (a large number are necessary to

Figure 12-9. *Radiant surface burner with high-efficiency ceramic radiant screen.* (Photo courtesy of Alzeta Corp., Santa Clara, CA, and AlliedSignal Composites, Inc., Newark, DE.)

produce enough electricity for a home); they only work when the sun is shining; and they produce direct current, whereas all of our household appliances use alternating current. But scientists are making progress. Costs are coming down, efficiency is increasing, batteries can store the electricity generated while the sun is shining for use at night, and equipment is readily available to convert from dc to ac. Don't be surprised to see a big increase in the use of solar cells in the future, with ceramics making possible the manufacture of the semiconductor silicon wafers.

Energy from Plant Matter. Another important renewable energy source is referred to by the general term *biomass*. Benefiting from biomass involves extracting energy from plants such as trees, grasses, and corn stalks. One way of extracting the energy is by burning the material and converting the resultant heat to electrical energy, just as in present electricity-generation stations that burn coal or oil. An alternative is to chemically convert biomass materials into a liquid fuel, such as alcohol, or a gaseous fuel, such as methane (natural gas), that can then be used to power your car or fuel your home furnace. Because of their high-temperature stability, corrosion resistance, and wear resistance, ceramics are important for both direct burning and chemical conversion methods of obtaining energy from biomass materials.

THE ROLE OF CERAMICS IN POWER GENERATION

The United States consumes staggering amounts of electrical power, and our needs are growing each year. Presently, those power needs are met primarily by a combination of sources, including large coal-fired and oil-fired plants, hydroelectric stations, and nuclear reactors. Smaller power sources include wind, geothermal energy, photovoltaic cells, gas turbine engines, and diesel engines. Fuel cells are an emerging technology that may be important in the future.

Power from Burning Nonrenewable Fossil Fuels

In Chapters 10 and 11, you learned about some of the ways ceramics are used in conventional coal-fired and oil-fired power-generation plants: as refractories, thermal insulation, thermocouple protection tubes, seals, valves, and wear-resistant linings. Ceramics also are important for pollution-control systems. One such example involves the use of high-temperature ceramic fibers woven into a fabric that can filter dust and unburned ash from hot waste gases before they exit a smokestack.

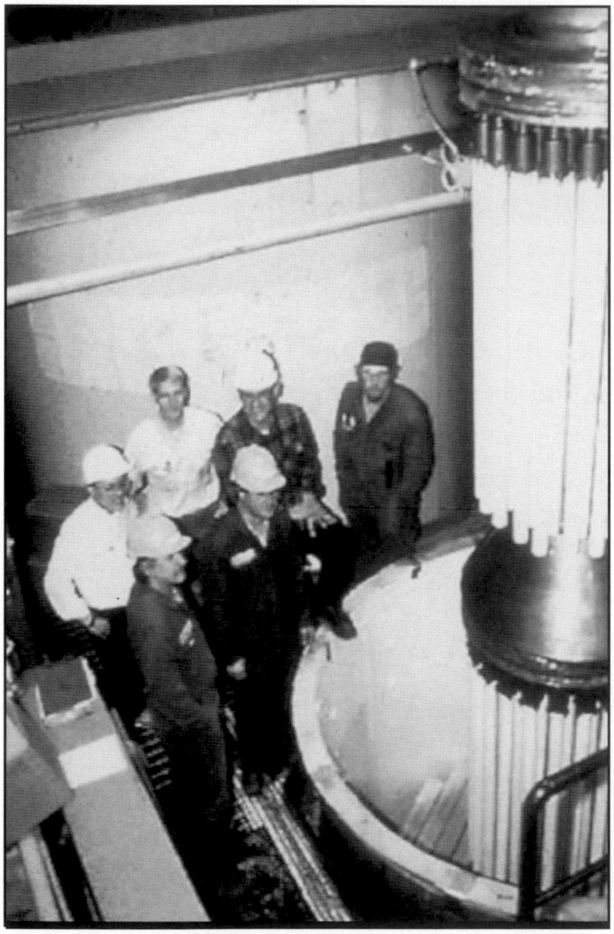

Figure 12-10. *Westinghouse candle filter system, showing the hanging porous ceramic tubes.* (Photo courtesy of Siemens Westinghouse, Pittsburgh, PA.)

Ceramics are even more important in new *combined cycle* coal-fired systems that have been under development during the 1980s and 1990s. These new systems, just now entering commercial service, produce about 35 percent more energy for every ton of coal than conventional coal-burning power stations. To understand where ceramics are important in these systems, you need to know a little about how a combined-cycle system works. A combined-cycle system is designed to run a power turbine (gas turbine engine) in addition to a conventional steam turbine, so that the combination uses the heat from the burning coal more efficiently. When the combined-cycle power station is running, powdered coal is mixed with air and burned in a large furnace chamber. This process produces hot gases (mostly air) that flow through the gas turbine engine rotors, causing them to rotate at high speed (as you learned in Chapter 5) to run an electrical generator. The hot gases leaving the turbine then pass over metal heat-exchanger tubes through which water is

Figure 12-11. *Composite ceramic candle filter tubes for hot-gas filtration.* (Photo courtesy of McDermot Technologies, Inc., Lynchburg, VA.)

flowing. The water boils to form high-pressure steam that then goes through a steam turbine to generate more electricity.

The problem with such systems is that coal contains a lot of impurities that don't burn cleanly, leaving a sticky ash material that gets carried along, suspended in the hot gases. If these ash particles go into a gas turbine engine, they cause severe erosion and corrosion or, even worse, they stick to the turbine parts and completely shut down the turbine. For a combined-cycle system to work, the ash particles must be effectively removed (filtered) from the hot gases before they reach the gas turbine engine. Metal filters can't withstand the temperature, and ceramic woven-fabric filters get clogged by the ash. A new type of ceramic filter, called a *candle filter,* had to be developed to solve the problem.

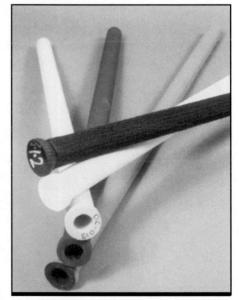

Figure 12-12. *Variety of ceramic candle filters.* (Photo courtesy of Siemens Westinghouse, Pittsburgh, PA.)

Figure 12-10 shows the inside of a hot-gas filter containing ceramic candle filters. The ceramic candle filters are the vertical tubes. Each of these hollow tubes is about 5 feet long, closed on one end, and very porous. The hot air containing coal ash particles flows around the outside of the porous tubes. The pores are large enough for the hot air to pass through to the inside of the tube, but not the ash particles. The ash builds up on the surface of each candle filter. Every few minutes, a blast of pressurized air is forced through the candle filter tubes from inside. This blast dislodges the layer of ash, which drops to the bottom of the filter chamber, where it can be removed easily as solid waste. The hot air that passes through the porous walls of the candle filter tubes is now free of ash and can be used in the gas turbine engine without fear of erosion or clogging (*fouling*).

Nuclear Energy

Ceramics are necessary in the nuclear power industry as fuel pellets, control rods, high reliability seals and valves, and a special means of containing (*encapsulating*) and stabilizing radioactive wastes for long periods of time. The fuel pellets used in a conventional reactor are made mostly of uranium oxide ceramic. A pellet only about three-eighths of an inch in diameter and one-half an inch long contains as much energy as 1 ton of coal, 150 gallons of oil, or 22,500 cubic feet of natural gas! The fuel pellets for a nuclear reactor are enclosed in rods over 12 feet long. A typical reactor contains over 45,000 fuel rods and can produce enough electricity to meet the needs of 350,000 all-electric homes.

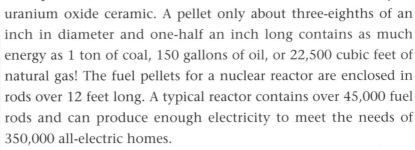

A uranium oxide/plutonium oxide ceramic fuel pellet measuring one-eighth of an inch in diameter and one-quarter of an inch long contains as much energy as 500 gallons of gasoline.

The nuclear reaction in uranium oxide fuel pellets produces heat, which is transferred to water to produce electricity by means of steam-driven generators. A more advanced reactor, called a *fast breeder reactor,* uses higher-energy fuel (a combination of uranium oxide and plutonium oxide) and transfers heat more efficiently and safely through a liquid metal. One of these ceramic fuel pellets measuring only an-eighth of an inch in diameter and one-fourth of an inch long has as much energy as 3 tons of coal, 12 barrels of oil, 500 gallons of gasoline, or 75,000 cubic feet of natural gas.

Perhaps the most important use of ceramics in the nuclear industry is for safely disposing of radioactive nuclear waste. The possibility that radioactive waste will escape or seep into the air or groundwater is a major fear for all of us. Ceramic engineers have devised ways to mix the radioactive waste with ceramic powders and fire this mixture at high temperature to form a solid glass or polycrystalline ceramic that's very stable. Fusing the radioactive waste into the very stable structure of the glass or other ceramic minimizes the chance of the radioactivity spreading into the surrounding environment.

Ceramics in Fuel Cells

As important as they are, coal-fired and oil-fired power plants aren't very efficient. Most of these power plants are able to convert only about 32 to 35 percent of the energy stored in the fuel to electrical energy. Advanced, combined-cycle coal-power plants may reach around 45 percent when they're optimized. We are wasting more than 50 percent of the energy contained in the fuel! An alternative that looks promising is *fuel cells*. Fuel cells make electrons available directly, through a chemical reaction, rather than by producing electricity indirectly by burning a fuel and converting the heat energy to electrical energy. Because fuel cells convert energy directly to electricity, less of the energy is wasted. A fuel cell can directly convert up to about 60 percent of the energy in the fuel to electricity and produce usable heat at the same time. If the heat also is used in some way (to heat a house or generate additional electricity), nearly 80 percent of the energy in the fuel can be tapped. That's double the efficiency of our present methods of power generation. Another benefit of fuel cells is that they give off much less pollution than fuel-burning processes.

There are many different types of fuel cells, but the one most important to the field of ceramics is the *solid oxide fuel cell* (SOFC). It operates similarly to the zirconium oxide (zirconia) oxygen sensor described in Chapter 9. If you remember, voltage is produced in the oxygen sensor if different oxygen concentrations are present on the inside and outside of the oxygen sensor tube. In the case of an SOFC, air is on one side of the zirconia and a fuel such as natural gas or hydrogen is on the other side. Oxygen from the air actually travels through the solid zirconia to react with the fuel and to produce electricity at the same time. Let's take a closer look at how this happens.

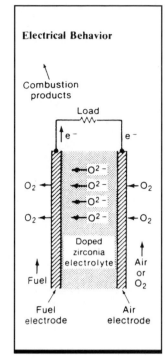

Figure 12-13. *Schematic showing the cross section of one type of solid oxide fuel cell.*

The zirconia is made into a thin sheet, tube, or membrane with no pores, so that air and fuel can't leak through. If the zirconia is coated on the two opposite surfaces with porous electrodes (electrically conductive material hooked up to an electrical circuit) and heated to about 1650°F, an amazing thing happens: Oxygen molecules from the air touch the electrode on the air side (which has a negative electrical charge) and latch onto two electrons from the electrical circuit, to form oxygen ions. These negatively charged oxygen ions are attracted to the other electrode (which has a positive charge) on the fuel side of the zirconia but must go through the solid zirconia to get there. The oxygen ions actually "sink" into the atomic structure of the zirconia and "swim" through the atomic structure, in the direction of the fuel and the positive electrode. When they reach the opposite surface, they give up their extra two electrons to the fuel-side electrode and

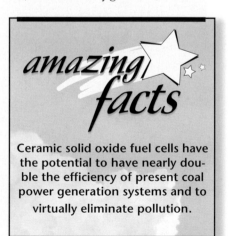

amazing facts

Ceramic solid oxide fuel cells have the potential to have nearly double the efficiency of present coal power generation systems and to virtually eliminate pollution.

react chemically with the fuel. The electrons now are available to flow through the electrical circuit to light a lamp or run a computer—or whatever else you choose to plug in.

Heat is produced by the chemical reaction between the fuel and oxygen, which keeps the fuel cell at an optimum working temperature. If hydrogen is the fuel, the only discharge from the fuel cell is water vapor (oxygen plus hydrogen equals water). If natural gas is the fuel, the SOFC discharges water vapor and carbon dioxide, essentially the same things we exhale when we breathe. Another advantage of the fuel cell is that it produces no noise.

The closest SOFC to commercialization appears to be a tubular configuration developed by Siemens Westinghouse Power Corporation. Having demonstrated small units for 8 years of continuous operation and a 25 kW unit for over 12,000 hours, Siemens Westinghouse plans to complete development of a megawatt-size SOFC early in the 21st century. This deadline coincides well with the deregulation of the power-generation utilities, an action expected to result in greater use of small, local power generators and less use of large, central power stations. As mentioned earlier, our present central power stations lose over 65 percent of the energy of the fuel when they convert it to electricity and then suffer further losses transporting the electricity across the countryside through power lines. Future power stations are striving for 45 to 55 percent efficiency, but that will be a big challenge. In contrast, the Siemens Westinghouse SOFC is expected to be 70 percent efficient and will suffer no losses from using power lines. Ceramic fuel cells have the potential to drastically improve the efficiency of electrical generation and the way we obtain electricity.

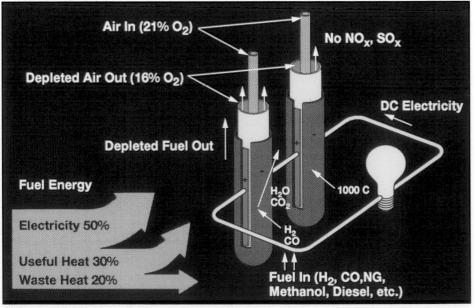

Figure 12-14. *Schematic illustrating the Westinghouse tubular fuel cell.* (Courtesy of Siemens Westinghouse, Pittsburgh, PA.)

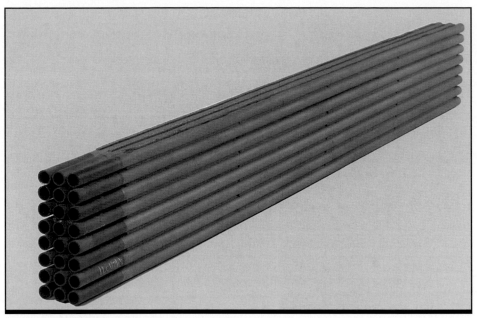

Figure 12-15. *Tubular solid oxide fuel cell module. (Photo courtesy of Siemens Westinghouse, Pittsburgh, PA.)*

Other SOFC designs being developed involve stacks of flat plates or honeycomb-like structures. These appear to hold the greatest short-term potential for applications requiring less than 100 kW, such as small buildings and homes. Very possibly, by 2010 or 2015 your new home will come with an SOFC that runs on natural gas or hydrogen and provides all of your electrical, heating, and cooling needs. These future SOFCs promise exciting benefits in energy efficiency, pollution reduction, and even the dramatic reduction of oil imports and our national trade deficit.

THE ROLE OF CERAMICS IN POLLUTION CONTROL

You've already learned about some ways in which ceramics control pollution. In Chapter 9, we reviewed how the automotive catalytic converter has reduced polluting emissions from automobiles by 1.5 billion tons since 1974. Earlier in this chapter, we discussed hot-gas filters for coal-fired power-generation plants and the fusing of radioactive nuclear waste into stable glass and polycrystalline ceramic compositions. In addition to these examples, any ceramic device or process that conserves energy or produces energy more efficiently than other devices reduces

Figure 12-16. *Experimental planar SOFC module. (Photo courtesy of SOFCo, a joint venture of McDermot Technologies, Inc., and Ceramatec, Inc., located in Salt Lake City, UT.)*

pollution. Thus, ceramic heat exchangers, improved thermal insulation, efficient lighting, igniters replacing pilot lights, higher-temperature efficient heat engines, and fuel cells all contribute to decreased pollution. But ceramics are also important in other areas, such as municipal garbage incineration, water and sewage treatment, hazardous waste incineration, and even oil-spill cleanup.

Incineration of Garbage and Hazardous Chemicals

One of our greatest pollution problems is the accumulation of the trash that we put out for pickup by the garbage truck each week. In the past, this trash was hauled to dumps or landfills. Over the years, we've made much progress toward recycling a few items, such as aluminum cans, newspapers, and some plastic, but there's still an enormous amount of solid waste. One technology introduced in recent years is municipal garbage incineration. The trash is taken to an industrial plant, where automated machines sort recyclable items that we missed (metal cans, a couple of different types of plastics, and glass) and burn the rest in a huge furnace (incinerator). The heat produced is used to generate electricity. Ceramics line the incinerators, provide thermal insulation to prevent unwanted heat loss, and provide chemical and thermal shielding for metallic heat-exchanger tubes that extract the heat for electrical generation. Ceramics also participate in cleaning ash particles from the air discharged by the incinerator.

Many materials are too hazardous to be disposed of in dumps or incinerated in municipal garbage incinerators. Examples are dioxins and other dangerous chemicals, as well as military chemical weapons. These materials are destroyed in special incinerators designed to assure that the dangerous compositions completely decompose to harmless materials. Ceramics also are required in these hazardous-waste incineration systems.

Water and Sewage Treatment

Most industrial processes require water, which often becomes contaminated with acids, bases, dissolved chemicals, and solid particles and must be treated before reuse or return to the environment. The channels for transporting the contaminated water, and the containers in which the water is treated, are often lined with ceramics. Sometimes the contaminated water is passed through porous ceramic filters that remove particles of unwanted materials. Frequently, gases are bubbled through porous ceramics into the water to cause chemical reactions that result in purification. Similar filtration and bubbler techniques are important in our cities' water and sewage treatment plants.

Containment of Oil Spills

One of the most surprising applications of ceramics is for containing oil spills from seagoing tankers. The idea behind containing an oil spill is to surround it with a floating structure called a *boom* and then eliminate the oil by burning it. Pioneered by Dome Petroleum in Canada, during oil exploration in the 1970s in the Canadian Beaufort Sea, the concept of burning an oil spill was first demonstrated with a stainless steel boom. The metal boom, however, was heavy and very expensive.

Shell Oil continued to work on metal booms in the early 1980s, for their exploration in the Alaskan Beaufort Sea; however, the most successful boom was mostly ceramic and was developed cooperatively by 3M (Minnesota Mining and Manufacturing) and American Marine. Called the *fireboom*, it consists of a string of floatable "logs" with a skirt draped below them and can easily be towed by boat to surround an oil spill. The logs are constructed of an inner core of a fused, porous, floatable ceramic wrapped with layers of stainless steel mesh and woven high-temperature ceramic cloth. The outer layer and skirt of the fireboom are made of a plastic called *polyvinyl chloride*. Once an oil spill has been surrounded by the fireboom, the oil is ignited and allowed to burn. Scientists have studied this contained burning and concluded that it's much kinder to the environment than allowing the oil spill to spread at sea or reach land.

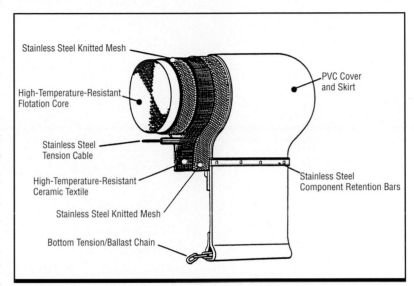

Figure 12-17. Schematic of the American Marine fireboom. (Courtesy of American Marine, Cocoa, FL.)

Each fireboom log is about 7 feet long and 12 inches in diameter, and the skirt extends below the logs about 30 inches into the water. A 50 foot section weighs only 425 pounds, many times less than a metal boom. During a test off the coast of Norway in 1988, 500 gallons of oil were successfully contained by a U-shaped boom towed by a boat. Ninety-five percent of the spill was burned in 30 minutes. The next year, a fireboom was used during a real emergency to contain and burn part of the oil spill from the Exxon Valdez, in Prince William Sound, Alaska. Approximately 15,000 to 30,000 gallons of oil were towed away from the main oil slick. About 98 percent of that oil was successfully burned in 45 minutes.

Figure 12-18. *Use of a ceramic fireboom to contain and burn an oil spill.* (Photo courtesy of American Marine, Cocoa, FL.)

OVERVIEW: TO A GREENER, HEALTHIER ENVIRONMENT

We've reached another major crossroad in civilization. We can go one direction—and continue to devour our natural resources at breakneck speed; or we can go the other direction and accelerate our efforts to change over to renewable resources. We can continue to pump pollution into our environment and take the chance of poisoning it beyond recovery, or we can accept the responsibility, as global citizens, to do our part to reduce pollution. In this chapter, we've explored some of the technologies that can make a difference, especially some ways that ceramics can help us conserve energy, generate electricity more efficiently, and reduce pollution. Each year, ceramics are becoming more important in these areas. "Green ceramics" are pieces in the technology puzzle that will enable us to decrease pollution and greenhouse gases in spite of increases in population, industrialization, and energy use. They'll help us continue to improve efficiency in our current energy-conversion processes and to make possible new and better processes.

CONCLUSION

Have ceramics reached their peak? *The answer is a resounding no! Ceramics will continue to grow in importance in nearly every other aspect of our modern civilization. Let's take a glimpse at things to come.*

In the future, personal computers will be nearly as powerful as today's supercomputers and will be linked to direct digital imaging and maybe even smart sensors for virtual-reality experiences and communications. This means that we'll be able to see people as we talk over the Internet and even will be able to "handle and feel" objects.

Tiny powerful microchips also will have a profound impact on medical technology. They will be linked to tiny sensors that will continually monitor our temperature, heart rate, and maybe even the amount of stress we are subjecting ourselves to during work or exercise. These microchips will give us early warning if a medical problem is developing and will probably be programmed to activate other implanted medical devices in the case of a life-threatening event. The IC chip in a heart-attack victim might signal changes to a pacemaker or activate a tiny defibrillator to release a burst of electricity from powerful miniature ceramic capacitors to get the heart beating again.

Future cancer treatment will involve trapping ceramic powders or beads inside cancerous tumors. When the ceramic is placed in a special energy field, its temperature will increase to a point that the tumor is destroyed without harming the body.

Other future medical miracles will include removing coatings that cause cardiovascular disease from the inside of blood vessels, live-cell encapsulation to allow genetically-engineered cells to be implanted into organs to manufacture things such as insulin, and improvements in imaging techniques and the electronics and computers that allow doctors to diagnose a problem more efficiently. Someday your physical exam will include a painless body scan in which a powerful computer will automatically search for any irregularities compared to previous scans and will provide an early warning of tumors, joint damage, and even kidney stones.

Energy will be a dominant issue as our world population continues to grow, especially as we deplete nonrenewable resources of oil and coal. Ceramics, because of their high-temperature capability, wear resistance, and corrosion resistance, will become more and more important. Less and less of our electrical energy will be generated at central power stations, and more electrical energy will be generated in our homes with ceramic fuel cells and solar cells, both linked to battery storage systems. Scientists and engineers will learn how to safely produce and use hydrogen, which will become an ultra-efficient and low-pollution fuel for home fuel cells as well as automobile fuel-cell engines.

Through government legislation in the United States and international agreements, there will be a worldwide effort to bring pollution under control. Ceramics will play a central role. Catalytic converters will be required on all motor vehicles worldwide and will be designed into other fuel-burning systems (fireplaces and heating stoves, for example) to reduce pollution. The ceramic heat exchangers that will be built into many of these fuel-burning systems will make efficient use of waste heat, which will reduce the amount of fuel used and decrease pollution. Ceramic hot gas filters will become widely used to remove particulate pollution from power generation stations, industrial processes, and waste incinerators. Other types of porous ceramic membranes will effectively remove chemicals and particles from fluids (especially water) to allow both the fluid and chemicals to be recycled. Even a complex process like papermaking will recycle nearly 100 percent of water and chemicals now used and will generate all the heat and electricity needed for the process.

Ceramic materials and other advanced materials have a bright future. I hope that this book has increased your awareness and appreciation of the importance of advanced materials in our modern world and in your personal world. I hope that you, too, agree now that ceramic materials are an amazing and often magical category of materials we cannot live without!

Index

A

acoustic waves, 154
acousto-optic ceramics, 81
active vibration control, 162
actuators, 159, 166
Advanced Research Project Agency, 95
agriculture, role of pottery in, 20
airbag sensors, 161
air-conditioner compressor seals, 197
aircraft, ceramics in, 112, 235–36
AiResearch Manufacturing Company, 95
air gap, 32
air-meter modules, 196
allografting, 174
alternate engines, 204, 269
alternating current, 127
alumina
    in armor, 259
    band gap of, 74
    in cutting tools, 241–42
    in experimental engine parts, 201–2
    in first high-performance ceramics,
        32–34
    in gemstones, 29, 30
    hardness of, 240
    in hybrid packages, 136–37
    in lasers, 79
    melting temperature of, 211
    in papermaking equipment, 250, 253
    in sodium vapor lamps, 267–68
    as substrate, 141
    transformation-toughened, 107
    use in aluminum smelting, 216
aluminum
    ceramics used in processing, 214,
        215–17
    ceramic tools for fabricating, 258
    melting temperature of, 114, 211
aluminum nitride, 141
aluminum oxide. See alumina
Alzeta, 272
amber, 126
Americas, early ceramic art in, 41–43
ammonium alum, 29–30
anodes, 216
arc tubes, 267–68
argon lasers, 80
armor, 258–59
artificial eyes, 177
artificial limbs, 177
Arts and Crafts movement, 51–52
atoms, properties of, 28, 62–64, 71
autoclaves, 104
autografting, 174
automotive ceramics
    in catalytic converters, 193–95
    and energy efficiency, 268–69
    in engine-control systems, 195–97
    experimental, 200–204
    in passenger compartment, 204–6
    in seals and rotors, 197–98
    in sensors, 192–93, 198, 220, 277
    in spark plugs, 31–33
auxiliary power units, 97
azurite, 21

B

ball and seat valves, 246, 247
band gap, 73, 126

band theory, 129
Bardeen, John, 129
barium titanate, 141–42, 151
baseball bats, 163
basic oxygen furnaces, 217–18
Bates, Wayne, 57, 58
bauxite processing, 214
Bayer, 33
bearings, 100–104
bending stress, 91–92
beryllium oxide, 141, 211
Binns, Charles Fergus, 56
bioactive ceramics, 171, 173
bioceramics, 171
Bioglass, 171, 175
biomass, 273
bisque-fired ceramic, 44
black pottery, 41
Blaschka, Leopold and Rudolf, 53–54
blast furnaces, 217
bleaching, of paper pulp, 252
Bloch, Felix, 129
block filters, 142–43
blood vessel blockages, 186
blow tanks, 250
boats, fiberglass, 13–14, 110–11
boat speedometers, 155
body panels, 207–8
bonds, atomic, 62–64
bone grafts, 174
bone restoration, 174, 175
booms, 281, 282
boron carbide, 259
boron nitride, 5
borosilicate glass, 240
Böttger, Johann Friedrich, 23, 49
boules, 29, 30
bowling balls, 14
braces, 172, 173
Brattain, Walter, 129
Bronze Age, 21, 38–39
bulletproof armor, 260
butterfly valves, 104, 245–46

C

cadmium sulfide, 74
camels, 43, 44
cameos, 49, 53
cam follower rollers, 199–200
cancer treatments, 182–85
candle filters, 274, 275
capacitors, 125, 141–42
carbon, in heaters, 229
carbonado, 244
carbon-carbon composites, 114
carbon dioxide, 216
carbon fiber composites, 111–12
cast glass, 27
casting, of metals, 222–24
catalysts, 193, 229–30
catalytic converters, 193–95
cathodes, 216
CAT scanners, 178–80
cellular phones, 142–43
cemented carbides, 240, 241, 242
centrifugal separators, 248
ceramic art
    in ancient times, 37–43
    during Middle Ages, 43–47
    modern expressions of, 53–59

post-Renaissance, 49–52
    during Renaissance, 47–49
ceramic-matrix composites, 112–14, 232–34.
        See also high-temperature com-
        posites
ceramics
    as art (see ceramic art)
    defining, 3
    effects of stress on, 90–92
    everyday uses of, 4–14
    importance in ancient world, 18–21
    internal structure of, 25
    medical applications for (see medical use
        of ceramics)
    modern, 28–34
    traditional, 22–27
ceramic sponges, 218–19
check valves, 246, 247
chemical composition, 28
chemical processing, 229–30
chemistry, and ceramic development, 28–29
China
    ceramic developments during Middle
        Ages, 43–44
    early ceramic art in, 40–41
    influence on European ceramics, 49
chippers, 250
Chou Dynasty, 22–23
chromium, 71, 79
clay, 19
Clean Air Act, 192
coal ash filters, 113–14, 274–75
coal processing, 248–49
cobalt oxide, 44, 71
coherent light, 79
coil-built pottery, 40–41
colon surgery, 182–83
color, 70–74
combined cycle systems, 274–75
combustors, 95
"comet" pattern pressed glass, 50
Compax, 244
complementary colors, 71
composites, fiber-reinforced. See fiber-rein-
        forced composites
compressive stress
    on bearings, 101
    in car windows, 204
    described, 90–91
computer disks, 258
concrete, 11, 112
conduction band, 73
conservation, 263–73
continuous fibers, 109–14. See also fiber-
        reinforced composites; high-
        strength ceramic fibers
copper-red color, 44
copper wire, 66–67
cordierite, 120, 195
core-formed glass containers, 26, 27
core-formed metals, 223
Corelle dishes, 7
Corning Glass Works, 116–18, 119
corundum, 29
cracking, in TTZ, 106
crossflow heat exchangers, 271
crowns, dental, 172
cryolite, 216
cubic boron nitride, 240
cubic zirconia, 65, 71–72

Curie, Pierre and Jacques, 150
current, controlling, 124–25
cutting tools, 98–100, 107, 226, 240–44
cylinder-in-cylinder valves, 246

D

Day, Delbert, 183
decorations. *See* ceramic art
decoys, piezoelectric, 154
dental ceramics, 171–73, 175
dental instruments, 103, 104
detached retinas, 176
dialysis, 185–86
diamond
    hardness of, 239–40
    industrial uses of, 242, 244
    refraction in, 64–65
    as substrate, 141
    synthetic, 243, 244
diamond coatings, 244, 260
dielectric ceramics, 141–43
diesel engines, 199, 202–4
digesters, 250
dimensional stability, 114–20
direct current, 127
direct-ignition modules, 196
dopants, 79
doping, 133
doppler sonar velocity systems, 155
double-paned glass, 265
down-hole systems, 256
drive shafts, 202, 207–8
Dubois de Chémant, Nicholas, 171–72
ductility, 90
dust pressing, 24

E

ear implants, 175
earthenware
    in ancient times, 19, 20, 22
    modern expressions in, 57–58
    See also pottery
Ebersole, William G., 244
Egypt, early ceramic art in, 37–38
electrical conductors, 125, 126
electrical insulators, 125, 126, 140–41
electric heaters, 229
electricity
    ceramics used in generating, 248–49,
        274–79
    development of ceramics for, 28
    history of development, 126–31
    properties of, 124
electroluminescent lighting, 4, 77–78, 206,
        207
electrolysis, 215
electrolytic cells, 215–16
electromagnetic radiation, and transparen-
        cy, 67–70
electromagnetic spectrum, 68
electromagnetic waves, 67
electromagnetic windows, 69
electromotive force (emf), 124
electron holes, 133
electronics
    future trends, 147
    history of development, 126–31
    integrated circuits, 130–39
    magnets in, 145–46
    other ceramic uses in, 140–45

overview of, 124–26
    recent developments in, 123
electrons
    band theory of, 129
    and color, 71, 73–74
    discovery of, 127
    and electronics, 124, 126
    and magnetism, 145
    and transparency, 62–64
electro-optic ceramics, 80–82
emulsions, ultrasonic mixers for, 158–59
enamels, 44
endoscopy, 180–81, 182–83
energy conservation, 263–73
engine-control systems, 195–97
experimental engine parts, 200–204
explosives detectors, 155–56
extrinsic semiconductors, 133
extrusion, 111, 225
eye repairs, 176–77

F

faience, 27
false teeth, 171–72
fast breeder reactors, 276
Ferguson, Dan and Nisha, 57, 58
ferrites, 146
fiberglass insulation, 10, 110, 264–65
fiberglass products, 110–11
fiber-optics
    for communications, 66–67, 81
    future applications for, 86
    medical uses of, 86, 181
    strength of material, 67, 89
fiber-reinforced composites
    benefits of, 110–12
    high-temperature, 112–14
    in prosthetic devices, 177
    in vehicles, 199, 207–8
    See also high-strength ceramic fibers;
            high-temperature composites
filters
    hot-gas filters, 114, 274
    for molten metals, 218–20
    for water and sewage treatment, 280
finishing, of metals, 213, 225–26
firebooms, 280, 281
firing, earliest evidence of, 19
fish finders, 155
flash-blindness goggles, 82
float-glass process, 85
floor tiles, early development of, 46, 48
fluorescence, 75
fluorescent lights, 75–76
foils, 252, 253
Fotoform process, 83, 84, 85
fuel cells, 276–79
fuel pellets, 276
fused quartz, 116

G

galena, 127
Galilei, Galileo, 115–16
gamma rays, 68
garbage incinerators, 280
Garcia, Tammy, 57
gas mileage, 268–69
gas turbine engines, 92–98, 112–13, 235–36
gate valves, 246
gems, synthetic. *See* synthetic gems

germanium, 129, 133
gibbsite, 214
glass
    art, 47–48, 53–55
    in automobiles, 204–5
    and benefits of transparency, 65–67
    early developments in, 27, 50
    effects of stress on, 90
    energy efficient, 265
    internal structure of, 24–25
    low-thermal-expansion, 115–20
    new manufacturing methods for, 85,
        228
glassblowing, 27, 40
glass-cloth lay-up technique, 110–11
glass microspheres, 183–85
glazes, 21, 39
golf equipment, 14, 108
grain boundary, 25
granite, 253
Gray, R. B., 151
The Great Dish, 52
Greece, early ceramic art in, 39–40

H

hafnium carbide, 211
Hale, George Ellery, 116, 118
Hall, Howard Tracy, 243
halogen lamps, 268
Han Dynasty, 22
hard ceramics
    for cutting, 98–100, 107, 226, 240–44
    for wear and corrosion resistance,
        245–58
hard ferrites, 146
hardness, 239–40
hazardous waste incinerators, 280
heart-valve implants, 176
heat and stretch therapy, 186–87
heat-exchanger process, 30–31
heat exchangers, 221, 230, 270–71
heat treatment of metals, 213, 226–27
Hench, Larry, 171
high-alumina ceramics, 32–34
high-pressure sodium vapor lamps, 266–68
high-strength ceramic fibers
    carbon fiber composites, 111–12
    development of, 108–11
    in high-temperature composites,
        112–14 (*see also* high-tempera-
        ture composites)
    in refractories, 231
    See also fiber-optics; fiber-reinforced
            composites
high-temperature composites
    aerospace applications of, 113, 114,
        232–36
    benefits of, 112–14
    for industrial processes, 113–14, 228–32
    use in metals processing, 212–27
    See also alumina; silicon carbide; silicon
            nitride; zirconia
high-temperature furnaces, 31, 33, 228–29
high-voltage wires, ceramic insulators for,
        140–41
hip replacements, 173–74
hot-gas filters, 113–14, 274–75
Hubble Space Telescope, 118–19
hybrid bearings, 102
hybrid ceramic electronics systems, 195–97

hybrid packages, 136–39
hybrid propulsion systems, 204, 269
hydroelectric dams, 272
hydrophones, 153

I

igniters, 165–66, 269–70
implants, 169–77
impurities, chemical discoveries about, 29
incandescent lights, 75–76, 266
incinerators, 280
incising, 38
indentation test, 239, 240
indium tin oxide, 77
Industrial Revolution, 51–52
infrared rays, 68
infrared sensing devices, 85
ingots, 218
ingrown toenails, 112
inlays, dental, 172
inspection, of metals, 157–58, 213, 226
insulation
    fiberglass, 10, 110, 264–65
    for refractories, 231
insulators, electrical, 28, 125, 140–41
integrated circuits
    development of, 130–31
    in engine-control systems, 195–97
    fabrication of, 133–36
    packaging of, 136–39
    structure of, 132–33
    vibration control systems for making, 162
intrinsic semiconductors, 132–33
investment casting, 222–23
ions, 215
iron oxide, 29, 71
iron smelting, 217–20
Ishtar Gate, 38, 39
Islamic ceramics, 45–47, 49

J

Jaffe, Bernard, 152
jet engines, 92–98
jewel bearings, 30
joint implants, 173–74

K

kaolin, 23
kidney dialysis, 185–86
kiln furniture, 227
kilns
    Bronze Age developments in, 21, 38–39
    development in China, 22
Kraft process, 249–55

L

lances, 218
lapping, 240–41
large-scale integration, 131
laser hosts, 79
lasers
    development of, 78–80
    and fiber-optic communications, 67
    medical uses of, 79–80, 176–77, 181
laser scanners, 12–13
lathes, ceramic cutting tools for, 99–100, 226, 241–42
lead, in early glazes, 39
lead zirconate titanate (PZT), 152
leaf springs, 208
lens replacements, 176

light
    and color, 70–74
    encoding information in, 66–67
    interaction with atoms, 63–64
    refraction of, 64–65
    visible, 68
light-emitting diodes (LEDs), 78
lighters, 165–66, 269–70
lighting, energy efficient, 266–68
lime kilns, 254–55
Lippershey, Hans, 115
liquid helium, 144
liquid nitrogen, 145
lithium sulfate, 153
live-cell encapsulation, 188, 189
liver cancer, 183, 184–85
load, ceramic resistance to, 91–92
Longshan pottery, 41
lost wax casting, 222–23
lusterware, 46

M

machine tools, 98–100, 102–3, 107, 226, 241–42
magnesium oxide, 211
magnetic moment, 145
magnetite, 145
magnets
    in automobiles, 205–6
    in electronics, 145–46
    household applications for, 8
Maiman, T. H., 78, 79
maiolica, 49
makeup, 5
malachite, 21
manifold air-pressure modules, 196
Marconi, Guglielmo, 128
master control modules, 195–96
matrixes, ceramic fibers in, 110–14. See also fiber-reinforced composites
medical use of ceramics
    in diagnostic equipment, 177–81
    for fiber-optics, 86, 181
    future applications, 189
    in laser surgery, 79–80, 176–77, 181
    for replacement and repair, 169–77
    for treatment and therapy, 181–88
Megadiamond, 244
Meissen artisans, 49
melting and holding of metals, 213, 220–22
melting temperatures, 114, 211
mercury, superconductive, 144
Mesopotamia, early ceramic art in, 37–38
metal halide lights, 268
metals
    ceramic-reinforced, 201–2
    effects of stress on, 90
    finishing, 213, 225–26
    heat treatment of, 213, 226–27
    inspecting, 157–58, 213, 225–26
    melting and holding, 213, 220–22
    melting temperatures of, 114, 211
    raw material preparation, 212–14
    role of ceramics in development of, 20–21
    shape forming, 213, 222–25
    smelting and refining, 213, 214–20
mica, 62
microphones, 151
microprocessors, 131. See also integrated circuits
microspheres, radioactive, 183–85

microstructure
    defined, 25
    transformation-toughened, 105–7
microwaves, 69
Middle Ages, ceramic development during, 43–47
middle-ear implants, 175
mine detectors, 155–56
Ming Dynasty, 44
mining, 248
mirrors, 205
missiles, 69
mixing equipment, 158–59
modern ceramics, brief history of, 28–34
molten-metal filters, 218–20
molybdenum disilicide, 229
monochromatic light, 79
motor vehicles. See automotive ceramics
Murano glass industry, 47–48
music, from piezoelectric ceramics, 164–65
Musrussu the dragon, 38, 39

N

nanometers, 67
natron, 27
natural gas igniters, 269–70
Newton, Isaac, 116
noisemakers, 164–65
nondestructive inspection devices, 157–58, 226
nozzle guide vanes, 95, 97
nuclear energy, 276
NZP, 120

O

ocean-floor mapping devices, 156–57
oil exploration, 156–57, 256–57
oil spill containment, 281, 282
opacity, 63–64
optical behavior, 61
    color, 70–74
    electro-optics, 80–82
    lasers, 78–80
    phosphorescence, 74–78
    photosensitivity, 82–85
    transparency, 62–70
optical fiber, 66–67, 89. See also fiber-optics
orbital implants, 177
ores, discovery of, 20
Organization of Petroleum Exporting Countries (OPEC), 263
orthodontics, 172, 173
oxygen sensors, 192–93, 220, 277

P

pacemakers, 187
Palomar Hale telescope, 116–18
papermaking, 249–55
PereoGlas, 175
petroleum processing, 229–30
petuntse, 23
phosphorescence, 74–78
phosphors, 75, 179
photochromic glass, 82–83
photodiodes, 179
photolithography, 135, 161–62
photonics, 81
photons, 63
photosensitive ceramics, 82–85
photovoltaic electricity, 74
physical therapy, ultrasound, 186–87

piezoelectricity
  in automobiles, 206
  everyday uses of, 4
  history of development, 150–52
  industrial uses of, 157–60
  medical uses of, 180, 186–87
  principles of, 125, 147
  in ultrasonic imaging, 180
  underwater applications of, 152–57
  in vibration detectors, 160–63
pill presses, 188
pilot lights, 269, 270
pingers, 154
pistons, ceramic-reinforced, 201
plastic encapsulation, 136
PLZT lenses, 82
polarization, 141, 142
pollution control devices
  in automobiles, 192–95
  for oil spills, 281, 282
  for power plants, 113–14, 274–75, 276
  for waste management, 280
polychromatic glasses, 84
polycrystalline ceramics
  diamond, 239, 244
  internal structure of, 25
  light transmission in, 64
  piezoelectric, 151–52
polyethylene plastic, 211
polymer-matrix composites, 110–12
polyps, 182–83
polyvinyl chloride, 280
porcelain
  discovery of, 23
  in early spark plugs, 32
  European development of, 49
  during Tang Dynasty, 44
pores, 25
positioners, 159
potter's wheel, 22
pottery
  development of decoration on, 37–43
  importance in ancient world, 18–21
  modern expressions in, 55–58
powder processing advances, 33
power generation, 248–49, 274–79
precious gems, synthetic. See synthetic gems
pressed glass, 50
press-molded stoneware, 56
press rolls, 253
Pro Osteon, 174
Prosser, Richard, 24
prosthetic devices, 177
pulp preparation, 250–52
pumps, 246, 247–48
Pyrex glass, 115, 116–18
pyrolytic carbon, 176
PZT, 152

Q

quartz crystals, 150, 151
quartz glass, 116
quartz watches, 164
quenching, 226

R

radar, 76
radiant burners, 113, 221, 271–72, 273
radiant surface burners, 228–29
radioactive glass microspheres, 183–85
radioactive waste, 276
radio communication, 128–29, 151

radio waves, 68, 69
radomes, 69
Rado watches, 6
rain-drop impacts, 69
recreational equipment, ceramics in, 13–14, 162–63
recycling, 254–55
reducing conditions, 41
refining, 213, 214–20
reflecting telescopes, 116–19
reflective coatings, 265
refraction, 64–65
refractive telescopes, 116
refractories, 214, 217, 231–32
renewable energy sources, 272–73
resistors, 125
respirators, 187, 188
Rochelle salt, 150, 151
rocket-nozzle liners, 235
Roman empire, 22, 40
rotary valves, 245
rotors, ceramic, 95, 197–98
rubies, synthetic, 12–13, 28, 30, 79–80. See also synthetic gems
ruby laser, 79

S

Santa Clara Pueblo pottery, 56, 57
sapphire, synthetic, 29, 30. See also synthetic gems
scalpels, 181–82
scintillators, 178, 179, 180
seal runners, 97
seals, 197, 245
seed crystals, 30
semiconductors
  coloring and, 72–74
  function of, 125, 126, 129
  in integrated circuits, 132–33
  light from, 78
  in solar cells, 74, 272–73
sensors, automotive, 192–93, 198, 220, 277
sgraffito, 47, 57
Shang Dynasty, 22–23
shape forming, of metals, 213, 222–25
Shockley, William, 130
silica. See silicon dioxide
silicon, 74, 133
silicon carbide
  art objects from, 57, 59
  for automotive seals, 197
  in ceramic-matrix composites, 112–13
  in experimental pistons, 201
  hardness of, 240
  in heaters, 229
  in heat exchangers, 271
  high-temperature stability of, 114
  for wafer boats, 135
silicon chips, 125. See also integrated circuits; semiconductors
silicon dioxide
  in halogen lamps, 268
  hardness of, 240
  melting temperature of, 211
  in Space Shuttle tiles, 234–35
silicon nitride
  for automotive seals, 197
  in bearings, 100–104
  in coal pipes, 249
  in diesel engines, 199–200
  discovery of, 92
  in gas turbine engines, 92–98, 236

for metal-cutting tools, 98–100, 107
  in papermaking equipment, 253
  for turbocharger rotors, 197–98
  in valves, 200–201, 246, 247
silver halide, 83
single-crystal growth, 30–31, 133–34
skin repair, 181
skis, vibration control systems for, 162–63
slab building, 56–57
slag, 214
slip, 137
slip casting, 44, 58
slurry, 137
smart contact sensors, 161
smart skis, 162–63
smelting, 213, 214–20
Snoeck, J. L., 145
soapstone, 21
sodium vapor lamps, 266–68
sodium zirconium phosphate (NZP), 120
soft ferrites, 146
solar cells, 74, 272–73
sol-gel silica microbeads, 188, 189
solid oxide fuel cells (SOFCs), 277–79
sonar, 76, 153–54
Space Shuttle, 69–70, 232–35
spark-control modules, 196
spark generators, 165, 166
spark plugs, 31–33, 236
sputtering, 135
stability, 114–20
stainless steel, 211
steel beams, inspecting, 157–58
steel manufacturing, 217–20, 246
Stone Age ceramics, 17, 18–19
stoneware, 22, 56–57
strength
  defined, 90
  high-strength ceramic fibers, 108–14
  and silicon nitride, 92–98
  and transformation-toughened ceramics, 104–8
stress, types of, 90–91
Subaru Telescope, 119
substrates, for integrated circuits, 135, 141
sucker rods, 256–57
suction box covers, 252, 253, 254
superalloys, 93, 114
superconductors, 125, 143–45
superhard abrasives, 242–44
surgery, ceramics in. See medical use of ceramics
synthetic gems
  coloring of, 71–72
  development of, 28, 29–31
  diamond, 243, 244
  in lasers, 79–80
  in scanner glass, 12–13
  use in jewelry, 58, 59

T

talc, 21
Tang Dynasty, 43–44
tape, 137–38
teeth, ceramic, 171–72
telephones, ceramics in, 142–43
telescopes, 115–19, 166
television, 76
temperature, and dimensional stability, 114–20
tensile stress, 91–94
textile industry, 255–56

thermal barrier coatings, 235–36
thermal expansion, 114
thermal shock, 93, 115
thermistors, 125, 198
Thompson, Joseph John, 127
thread guides, 255–56
tiles
    early development of, 23–24
    in rock crushing equipment, 212
    on Space Shuttle, 232–35
timing plungers, 199
tin oxide, 39
titanium carbide, 240, 250
toughness
    defined, 90, 104–5
    of diamond, 244
    of latest silicon nitride, 96
tourmaline, 150
Tower of Babel, 39
traditional ceramics, brief history of, 22–27
transducers
    in ultrasonic imaging, 180
    in underwater devices, 154–57
    use in dialysis, 186
    use in industry, 157–58
transfer printing, 24
transformation-toughened ceramics
    alumina, 107 (see also alumina)
    in diesel engines, 199
    in joint implants, 173
    in metal extrusion, 225
    new applications of, 107–8
    in papermaking equipment, 253
    zirconia, 104–8 (see also zirconia)
transformers, 160
transistors, 129–30
transparency
    benefits of, 65–67
    and electromagnetic radiation, 67–70
    elements of, 62–64
travel, role of pottery in, 20
Tuc d'Audobert Cave, 19
tundish nozzles, 224
tungsten carbide-cobalt cermets, 240, 241,
    242
tungsten filaments, 266
turbocharger rotors, 197–98
turboprop engines, 94, 95
two-chamber kilns, 21, 38–39

U

ultra-low-expansion (ULE) glass, 119
ultrasonic cleaners, 158
ultrasonic imaging, 180
ultrasonic machining equipment, 159–60
ultrasonic nondestructive inspection, 158
ultrasonic therapy, 186–87
ultraviolet rays, 68, 75–78, 84
underwater devices, 152–57
uranium oxide fuel pellets, 276

V

V-22 vertical takeoff aircraft, 111–12
vacuum tubes, 128–29
valence electrons, 73–74
valves, ceramic, 200–201, 245–46
varistors, 125, 198
veneers, dental, 172
Venetian glass, 47–48
Verneuil process, 29, 30
very large-scale integration, 131
vibration detection and prevention, 160–63

Victorian Age, 53
visible light, 68, 70–71
voltage, 124–25
voltage-regulator modules, 196
von Tschirnhausen, Ehrenfried Walther, 23,
    49

W

wafer boats, 134, 135
Ware Collection of Blaschka Glass Models of
    Plants, 54
waste management, 280
water and sewage treatment, 280
waterjet cutting, 241
water-pump seals, 197
wavelength, 67, 70–74
Wedgwood, Josiah, 23, 49, 172
wheel balancers, 161
whiskers, 109, 201
white light, 70
wind energy, 272
windows, 204–5, 265
wire-drawing tooling, 257
wood ashes, 27
woodyards, 249–50
workplace, ceramics in, 11–12

X

X rays, 178–79

Z

zirconia
    in diesel engines, 199, 203–4
    in fuel cells, 277
    hardness of, 240
    in heaters, 229
    high-temperature stability of, 114
    in joint implants, 173
    in metal extrusion, 225
    in oxygen sensors, 192–93
    in papermaking equipment, 253
    in thermal barrier coatings, 235–36
    transformation-toughened, 104–8
zirconium oxide. See zirconia